SpringerBriefs in Energy

Computational Modeling of Energy Systems

Series Editors

Thomas Nagel, Research GmbH - UFZ, Helmholtz Centre for Environmental Research GmbH - UFZ, Leipzig, Germany

Haibing Shao, Helmholtz Centre for Environmental Resea, Department Environmental Informatics Helmholtz Centre for Environmental Resea, Leipzig, Germany

Computer aided engineering has become indispensable in all major engineering sectors to shorten development times, limit expensive testing, and understand technological working principles as well as failure mechanisms.

In science, virtual in silico laboratories play a very similar role and have become an important means to testing research hypotheses and running virtual experiments. Many of the world's societies are currently transforming their energy systems to renewable energy sources, decentralized energy landscapes, and smart grids. The success of this transition is vital for the establishment of an ecologically, economically and socially sustainable future. The details of the future energy mix remain to be established and science must provide options for future energy systems along with the means to assess these options. Computer aided engineering and assessment capabilities for the novel components of those systems and the systems themselves need to be developed and new generations of engineers and scientist trained to use them effectively.

This subseries puts a spotlight on advanced computational and theoretical methods, tools, and frameworks for the design, analysis, optimization, and assessment of a diverse range of energy technologies and systems. The intention is to make the methods transparent and to allow engineers and scientists from different disciplines to enter the field of energy research enabling them to perform meaningful simulations for the advancement of clean and secure energy systems.

Sabrieh Choobkar • Seyed Mohsen Hashemi

Digital Twin Technology for Smart Grid

Springer

Sabrieh Choobkar
Information and Communication
Technology Research Group
Niroo Research Institute
Tehran, Iran

Seyed Mohsen Hashemi
Power System Operation and Planning
Research Group
Niroo Research Institute
Tehran, Iran

ISSN 2191-5520　　　　　　　　ISSN 2191-5539　(electronic)
SpringerBriefs in Energy
ISSN 2570-1339　　　　　　　　ISSN 2570-1347　(electronic)
Computational Modeling of Energy Systems
ISBN 978-3-031-90098-3　　　　ISBN 978-3-031-90099-0　(eBook)
https://doi.org/10.1007/978-3-031-90099-0

© The Editor(s) (if applicable) and The Author(s), under exclusive license to Springer Nature Switzerland AG 2025

This work is subject to copyright. All rights are solely and exclusively licensed by the Publisher, whether the whole or part of the material is concerned, specifically the rights of translation, reprinting, reuse of illustrations, recitation, broadcasting, reproduction on microfilms or in any other physical way, and transmission or information storage and retrieval, electronic adaptation, computer software, or by similar or dissimilar methodology now known or hereafter developed.
The use of general descriptive names, registered names, trademarks, service marks, etc. in this publication does not imply, even in the absence of a specific statement, that such names are exempt from the relevant protective laws and regulations and therefore free for general use.
The publisher, the authors and the editors are safe to assume that the advice and information in this book are believed to be true and accurate at the date of publication. Neither the publisher nor the authors or the editors give a warranty, expressed or implied, with respect to the material contained herein or for any errors or omissions that may have been made. The publisher remains neutral with regard to jurisdictional claims in published maps and institutional affiliations.

This Springer imprint is published by the registered company Springer Nature Switzerland AG
The registered company address is: Gewerbestrasse 11, 6330 Cham, Switzerland

If disposing of this product, please recycle the paper.

Preface

Computational methods and algebraic algorithms have been improved drastically in the last two decades, boosting up information and digital technologies (IDTs). This digital transformation in Industry 4.0 revolution integrates physical and computational components and forms cyber-physical systems. Many industrial fields could contemplate such fast revolution and soon recognized the urgency of IDT adoption in their machinery and elevate automation and computer-aid processes. The growth of advanced technologies has changed lifestyle drastically and is believed to make complex tasks more straightforward and more effective.

The digitalization era proceeds to unique technologies, such as cellular communication networks, Internet of Things, artificial intelligence, and virtual and augmented reality, resulting in pervasive smartness and intelligence. However, these recent technologies have partly answered system requirements of connection, data gathering, analysis, and visualization. Therefore, a comprehensive view of the entire system's operation and performance is still needed.

One of the recent technological advancements is the concept of "digital twin (DT)." This rapidly growing technology is already being utilized in modern industries, such as manufacturing and aerospace.

A DT is a virtual replica of an object or system that accurately mirrors its characteristics and behaviors. This innovative technology does not focus on a single approach or a specific technology but combines multiple trending technologies to address complex challenges of a system. This structure makes DT an advanced tool for meeting the intricate requirements of future applications.

Among dominant and progressive industries, smart power grid is a prime candidate for deploying DT technology. Smart grid (SG) system is a complex and unpredictable network with thousands of controllable devices. Although SG has implemented monitoring and control systems, there is still a considerable gap between an utter system showing the behavior of small parts and an entire structure. DT grants a complete view of a whole system and its subset segments and can focus on parameter changes and their effects in SG applications. Another significant point is harmful and costly faults and errors in SG. A custom-tailored DT able to predict and elucidate faults has a high priority to be deployed in SG domains.

This book offers a comprehensive overview of digital twin (DT) for electrical engineers. It covers all the essential knowledge about the concept, applications, and building technologies of DT. This book then reviews the structure and subsystems of SG while introducing DT novelties for each section of this modern power system.

The main objectives behind writing this book are as follows:

1. To introduce and express fundamental concepts of DT technology, as well as its structure and building modules.
2. To explore the applications of DT in industry, highlighting their benefits, consequences, and anticipated challenges.
3. To study and explore all aspects and advantages of DT implementation in confluence with SG and its subsystems.
4. To illuminate how to draw a roadmap of DT technology in SG.
5. To commence a focused study for further research on DT technologies and SG.

Tehran, Iran Sabrieh Choobkar
Tehran, Iran Seyed Mohsen Hashemi

Contents

1 Digital Twin Overview 1
2 Digital Twin in Industry 9
3 Digital Twin: A Novel Solution for Smart Grid Applications 15
4 Digital Twin for Smart Power Generation 21
5 Digital Twin for Smart Power Distribution 29
6 Digital Twin for Smart Power Transmission 39
7 Digital Twin for Future Smart Grid 43
8 Conclusions 47
References 49
Index 53

List of Figures

Fig. 1.1	Digital twin concept	2
Fig. 1.2	DT systematic structure	4
Fig. 2.1	Applications of DT	10
Fig. 3.1	The effect of smart grid concept on different sections of the power system	16
Fig. 3.2	Smart grid structure	17
Fig. 4.1	Smart power generation subsystems	22
Fig. 4.2	DT advantages to PV systems	24
Fig. 4.3	DT solutions for wind turbines	26
Fig. 4.4	DT opportunities for energy storage systems	28
Fig. 5.1	Smart power distribution subsystems	30
Fig. 5.2	DT and AMI cooperative system	32
Fig. 5.3	DT image of MG and its opportunities	35
Fig. 5.4	DT advantages for EV	37
Fig. 6.1	DT for smart transmission	40
Fig. 7.1	Future of smart grid	44
Fig. 7.2	DT effects on SG main subsystems	45

List of Tables

Table 1.1 Lexical definitions of digital twin..3
Table 1.2 Dimension-based definitions of digital twin 4

Abbreviations

AI	Artificial intelligence
AMI	Advanced Metering Infrastructure
AR	Augmented reality
BMS	Battery management system
CAES	Compressed air energy storage
CAV	Connected autonomous vehicles
CCHP	Combined cooling, heating and power
CDT	Cognitive digital twin
CHP	Combined heat and power
CPP	Critical peak pricing
DA	Distribution automation
DER	Distributed energy resources
DFIG	Doubly fed induction generator
DLC	Direct load control
DR	Demand response
DSM	Demand side management
DSO	Distribution system operator
DT	Digital twin
DTaaS	Digital twin-as-a-service
EDRP	Emergency Demand Response Programs
EMS	Energy management system
EV	Electric vehicle
FACTS	Flexible AC transmission systems
FLISR	Fault location, isolation, and service restoration
GE	General electric
GPS	Global positioning system
ICT	Information and communication technologies
IDT	Information and digital technologies
IGBT	Insulated gate bipolar transistor
IIoT	Industrial Internet of Things
ILP	Interruptible load programs

IoT	Internet of Things
MG	Microgrid
ML	Machine learning
O&M	Operation and maintenance
P2P	Peer-to-peer
PLC	Power line communication
PMU	Phasor measurement unit
PSH	Pumped-storage hydropower
SaaS	Software-as-a-service
SDO	Standard Development Organizations
SG	Smart grid
SOC	State of charge
STATCOM	Static synchronous compensator
SVC	Static VAR compensator
TCSC	Thyristor-controlled series capacitor
UPFC	Unified power flow controller
VPP	Virtual power plant
VR	Virtual reality
WAMPAC	Wide area measurement protection and control
WLAN	Wireless local area network

Chapter 1
Digital Twin Overview

Data in recent technologies is a key enabler in transferring basic, valuable, and analytical information. It feeds data-driven science, which involves inspecting, transforming, and modeling data with diverse investigative methods to extract meaningful information for different purposes. Data science has brightened the perspective on smart life and accelerated the development of smart cities and intelligent industries.

While fast processors and high-resolution displays are available in computer systems, acquiring valid sources of information and data model results shapes visual figures that represent simulation. Simulation techniques have been of significant interest for analyzing input data from multiple sources and gaining insight into the behavior of a relatively simple system without testing the exact system in reality. Later, co-simulation ideas suggested coupling separate simulators to make more powerful computer results that are able to emulate semi-complex operations in a virtual environment.

Remember that DT is more than just illustrating simulation results; it is a factual equivalent in the virtual world. Visually demonstrating simulation outcomes in displays could show parts of what happens in the study system. Most industries require a more robust copy that perfectly analyzes and reflects the actual object or system, delivering almost all its aspects, dimensions, and features.

Digital twin (DT) is defined as the digital replica of a physical asset that mimics its behavior in real time. Leveraging multiple data sources, the innovative concept of DT seeks to provide comprehensive analytical insights from past, current, and probable future data origins. The DT accurately reflects its physical twin's functions and operations, interacting with its parent object throughout its lifecycle. Analyzing data within DT yields an intelligent structure that optimizes, predicts, and maintains the actual object's functionality.

Essentially, DT architecture consists of transformative technologies that come together. A component-based view of a DT system is shown in Fig. 1.1 which consists a Siemens W501F Gas Turbine. Here, DT is introduced as a combination of three main components: physical object, communication medium, and virtual entity.

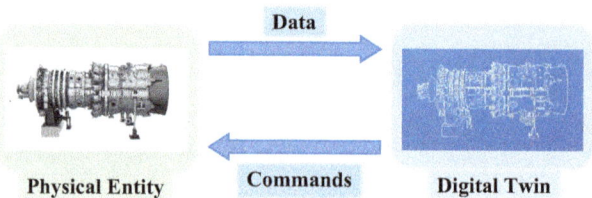

Fig. 1.1 Digital twin concept

The origin of DT dates back to 2002, when Dr. Michael Grieves first proposed the concept. He defined DTs in three modules: a physical entity, a digital counterpart, and a connection interface between the two [1]. This is illustrated in Fig. 1.1. Later, in 2010, NASA expanded upon this idea and applied it to complex space vehicles. The DT design in NASA inspired researchers to refine further and extend the definition to a more precise and accurate concept. The image of DT was developed and promoted until Gartner Inc. listed DT among the top 10 strategic and trending technologies in 2017 and 2018. Since then, several standard development organizations (SDOs) have been studying DT to release standards necessary for its implementation and development in various potential applications.

1 Definition and Architecture of DT

Since its inception, DT has garnered significant attention. The concept of DT involves creating a real-time mirror of an object or system. At first glance, it may seem straightforward, but it involves a complex structure that integrates advanced ideas and techniques. DT is a digital entity capable of reflecting the behavior of its physical counterpart in real time and with total accuracy.

A comprehensive definition of the DT concept should encompass the following characteristics:

- A DT is expected to reflect the "real-life" behavior of the physical object or system, including all its features and characteristics.
- DT represents a real-time, three-dimensional (3D) representation of the actual object, considering all relevant details based on aggregated data.
- DT is a twin that is easy to interact with. The real and virtual sides engage in bilateral data communication, creating a closed-loop system.
- DT is a data-driven technology that leverages diverse data sources to supply analytics and modeling modules.
- DT combines several new technologies to shape a dual object that is as similar to the real twin.

Investigating the literature reveals several definitions for a DT, focusing on mapping physical facility behavior into its virtual copy. Unfortunately, there is no

1 Definition and Architecture of DT

commonly accepted definition for DT [2, 3], primarily because this concept lacks essential and authoritative fundamentals a novel technology requires.

It is essential to discover and classify current research to reconcile academic and industrial perspectives. Several DT publications have employed lexical analysis but lacked a semantic view on extracting clear insights from the DT concept. Table 1.1 presents some lexical definitions of DT, chosen to illustrate the evolution of insights into the DT concept and to cover ideas about DT from different perspectives.

Different production-based DTs employ a dimension-based approach to recognize their concepts through features and dimensions. This approach highlights well-known dimensions of DT and uses them to explain a substantial part of this technology (see Table 1.2).

The two abovementioned groups of DT definitions brighten up mindsets to distinguish the general concept. However, DT brings multiple terminologies and challenges that require extended descriptions. To view a DT more broadly, we introduce a systemic view of DT where the whole scheme is considered as interrelated subsystems. This breaks the limited boundaries of physical and virtual worlds and lets the DT context be studied deeply in layers. In Fig. 1.2, we suggest a conceptual view of DT in which DT is defined as an integration of model, data, and functions overlapping at least two physical or functional domains.

To investigate technical aspects of DT, a systemic view is necessary, as depicted in Fig. 1.2. To provide detailed information about the real object's components, a sensing system comprising sensors and actuators is designed and installed on the physical object. This sensing package includes bidirectional modules that transmit and receive data and commands, effectively creating a comprehensive IoT system.

DT functionality relies on data management middleware, which enables the IoT system and other data resources to feed a data concentrator, potentially located in a cloud center. This data center aggregates and stores data, analyzes and models it

Table 1.1 Lexical definitions of digital twin

	Definitions of digital twin	Year
1	An integrated multiphysics, multiscale, probabilistic simulation of an as-built vehicle or system that uses the best available physical models, sensor updates, fleet history, etc., to mirror the life of its corresponding flying twin [4]	2012
2	A coupled model of a real machine that operates in the cloud platform and simulates health condition with an integrated knowledge from both data-driven analytical algorithms as well as other available physical knowledge [5]	2013
3	A set of virtual information constructs that fully describes a potential or actual physical manufactured product from the micro-atomic level to the macro geometrical level. At its optimum, any information that could be obtained from inspecting a physical manufactured product can be obtained from its digital twin [6]	2016
4	Simulation toward real-time control and optimization of products and production systems [7]	2017
5	A real mapping of all components in the product life cycle using physical data, virtual data, and interaction data between them [8]	2018
6	A digital twin is a virtual model that replicates a physical system or process, providing real-time data and feedback on its performance [9]	2023

Table 1.2 Dimension-based definitions of digital twin

	Dimensions	Definitions of digital twin	Year
1	Three	DT is categorized based on three dimensions: hierarchical levels, lifecycle phases, and common uses. It then allocates a DT within a cubical structure. However, it does not clarify a definition or some criteria for a virtual product or a DT [3]	2020
2	Three	Terminologies called: DT prototype (contains physical object's information which mirrors its virtual twin), DT instance (actual authentic product), and DT environment (a multiple virtual space for operating on DT) [6]	2017
3	Five	Five layers of cyber-physical data store layer, primary processing layer, models and algorithms layer, analysis layer, and visualization and user interface layer. However, these layers are extracted from the fundamental process of achieving a DT [10]	2019
4	Five	The model is formulated with parameters of physical entities, virtual models, services, DT data, and connections [11]	2019
5	Eight	The analysis shows behaviors and the context of a DT. These dimensions are categorized as integration breadth, connectivity modes, update frequency, CPS intelligence, simulation capabilities, digital model richness, human interaction, and product lifecycle [12]	2019

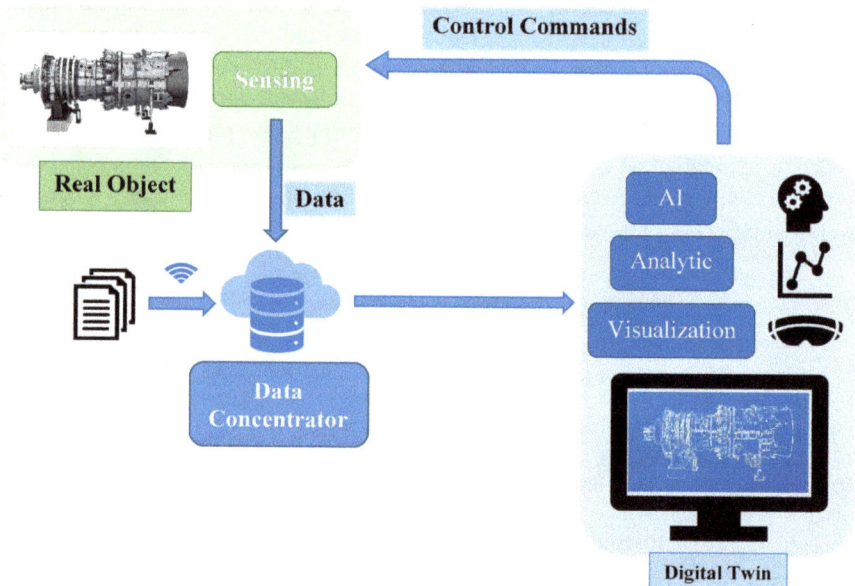

Fig. 1.2 DT systematic structure

using big data techniques, and then passes the outcomes to the digital twin peer. Technology modules in the digital counterpart comprise modeling, analysis, and simulation to extract and visualize advanced results for users.

2 Digital Twin Building Technologies

Based on the above descriptions, DT is constructed with prominent cutting-edge technologies. Hereunder, the main components and enabling technologies of DT are explained.

2.1 Cloud Computing

This technology efficiently and effectively assigns and shares vast computing resources (servers, storage units, services, etc.). On-demand access to cloud infrastructure provides security, cost-effectiveness and convenience for proactive and aggregated control and management in various applications.

In DT design, the cloud serves as a robust platform, offering reliable and sustainable servers, computing resources, along with large storage units. Since DT requires lots of real-time processing, a considerable solution is to transfer all information to a powerful cloud platform capable of handling large volumes of data. Additionally, cloud computing provides data storage, rapid data processing, and data modeling capabilities. Within a cloud-based DT, unique identity numbers are assigned to both the physical object and its virtual twin. In this context, the cloud acts as a bridge between the physical and application layers.

2.2 Internet of Things (IoT)

IoT defines frameworks and standards for connecting numerous devices. This approach involves assigning a unique identifier (ID) and internet protocol (IP) address to each physical object. IoT encompasses specialized platforms for data transfer, information storage, and sometimes, data analysis. Furthermore, industrial IoT (IIoT) is more prevalent in industrial settings, aiming to enhance productivity in various industries.

DT requires multiple connections and significant number of real-time data transfers, which can be established through IoT schemes. IIoT platforms enable connections between hardware components (e.g., sensors, cameras, actuators, and devices) and transfer their data to and from virtual models. By leveraging connected devices, it facilitates task evaluation and real-time responses.

2.3 Data-Driven Techniques/Big Data

Different data-driven strategies interpret factual data to derive information and possible insights, leading to more efficient solutions. Various methods of statistics and machine learning (such as deep learning) use complex algorithms to draw out observations from collected data. This contributes to management revolution and performance optimization. The focus of big data techniques includes aggregation, storage, pre-processing, formation and remediation, management, analysis, and utilizing massive amounts of various and complicated data.

Since DT is a data-driven technology, it strongly relies on data structures. Therefore, incorporating big data analysis into DT will extract valuable information, provide insightful management, and lead to more precise decisions [13].

2.4 Modeling and Simulation

A model represents a system structure with the aim of apprehension and analysis of its performance. A high-fidelity model provides details of system operation and gives a clear opinion of how the system works. Simulation is a process of studying a model to recognize and derive the behavior of a system. It is a tool used to evaluate the built model and extract operational characteristics of the object.

In a DT structure, modeling and simulation are crucial components of creating a comprehensive representation of the physical object, combining factual physical phenomena with its virtual counterpart. A computer model mirrors behavior of the physical object, while DT also tracks its real-time actions to refine its digital replica. A DT involves high-level mathematical modeling that addresses optimization problems in manufacturing execution. Furthermore, simulation results derived from the model must be updated to reflect the object's activities.

2.5 Artificial Intelligence (AI)

AI technology has become a prominent trend in industrial research centers, offering distinct value propositions. An intelligent system is capable of perceiving its environment and reacting in an intelligent manner. Currently, various AI techniques are gaining interest, including deep learning, statistical methods, and computing intelligence.

Advanced AI methods support decision-making in DT by providing insightful results from their analyses. Here, AI learning techniques empower DT to analyze actual data and establish intelligent decisions, rather than simulating output data. They also simplify complexity of DT applications, enhance the physical object's functionality, and increase design flexibility for intelligent agents to have more decision-making options in the context of DT applications. In these systems, a specific type of intelligent agent is required, known as a cognitive engine.

2.6 Visualization, Virtual and Augmented Reality Technologies

These topics introduce human interface techniques with various applications in large industries such as gaming, education, and healthcare. They are responsible for making analysis results and details visible on a defined dashboard. Visualization, in general, refers to a group of techniques that project a process and its physical features. VR demonstrates real-world information in a virtual form. AR is another visualizing technology that mixes virtual and physical environments.

To effectively interact with digital content, DT applications leverage several capabilities of VR and AR. Specifically, entities in multiscale simulations, which are central to DT, necessitate VR and AR models to accurately reflect visual and geometric characteristics.

2.7 Communication Systems

Fast data transfer and reliable communication networks are essential foundations for building any smart system. The deployment of recent cellular communication networks (4G and 5G) has brought numerous significant benefits, including high data rates, short delays, and high reliability. Additionally, 5G networks can provide customized service slices tailored to specific application requirements.

A DT comprises multiple connections between physical and virtual systems. To function effectively, it requires a communication system that offers high quality of service (QoS), ultra-high data rates, very low latency, simultaneous data transfer between data sources, and ultra-low power consumption. 5G technology meets the requirements for a DT design and makes a real-time data communication achievable.

2.8 Security

A DT introduces simulation, monitoring, optimization, and prediction of a physical system, making it vulnerable to attacks. Therefore, security of a DT should be considered at the device and equipment levels, smart connections, platform, and application services. Furthermore, a prospective DT will be an autonomous virtual entity, accompanied by security challenges. Contrary methods against threats and obstructive activities in autonomous systems could be the required solutions in a DT design.

Chapter 2
Digital Twin in Industry

Technologies have long been intertwined with industrial sectors. Automation represents the technological intervention in various industrial ecosystems, spanning from product design to manufacturing processes. Recently, advanced technologies such as IIoT, data science, and AI have been combined into automation to optimize processes and reduce costs, ultimately leading to the development of smart industries. Notably, the impacts are reciprocal, as technologies are refined and strengthened by industry research groups.

In technology companies, IBM's definition of DT emphasizes the importance of real-time data throughout a product's lifecycle. Moreover, to describe the viewpoint of energy industries on the concept of DT, some of their definitions are extracted as follows:

- ABB pictures a DT as an evolving digital profile of the historical and current behavior of a physical object or process that helps optimize business performance and is based on massive, cumulative, real-time, real-world data measurements [14].
- GE defines DT as a structured collection of physics-based methods and advanced analytics that models the current state of every asset within a digital power plant [15].
- Siemens Company defines a DT as a virtual double of a product, a machine, a process, or a complete production facility. It contains all data and simulation models relevant to its original [16].
- EPRI, the American Electric Power Research Institute, defines a DT as a digital replica of physical and functional characteristics of assets, providing information to systems or personnel to inform tactical or strategic operational decisions [17].

The industrial objectives of a DT recount that it must support a variety of different functions in complex systems and structures [18], such as:

1. Asset investigation and status analysis: including system anomalies monitoring, monitoring of asset material for any possible deformations, and asset reliability estimation.
2. Digital reflection of an asset's behavior: it involves analyzing long-term data related to a system's life cycle, predicting performance in different environments, and creating a virtual copy of the asset.
3. Asset management: optimal asset management involves the interaction between components and modules, statistical analysis for informed decision-making, system optimization, and life cycle optimization utilizing past and future data. Additionally, predicting future conditions is crucial.

1 Applications of Digital Twin in Industry

DT concept introduces a new approach that shows considerable potential with improved solutions to several industries. From aerospace to connected autonomous vehicles (CAVs), from healthcare sectors to intelligent farming, DT transforms industries by offering optimized decisions in management, control and monitoring, malfunction analysis and maintenance, and machinery safety and industrial security. Figure 2.1 highlights the applications of DT, emphasizing their potential in smart power grid systems.

The need for DT in industry is becoming increasingly pressing, and complex manufacturing companies cannot afford to ignore it. DT is of significant interest because it has the capability to organize information and provide a comprehensive solution to both targeted and non-targeted challenges. Depending on the approach and complexity chosen for DT, it improves manufacturing and optimizes production lines in the three phases of design, production, and maintenance.

Despite being in its early stages in industry applications, DT has the potential to simulate complex operations, provide enhanced visibility, develop innovative products, empower automation, and offer better insights into management processes and decision-making.

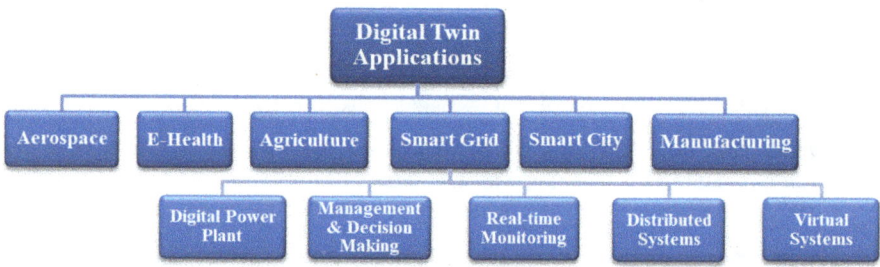

Fig. 2.1 Applications of DT

It is worth noting that DT can be designed and run at different scales, such as for various parts or sections of a complex asset like a gas turbine. The next step is to integrate DTs and form a comprehensive networked-DT (also referred to as a connected-DT) for the property.

Furthermore, DT in industry can be a product whose model and platform operate in the cloud. This type is offered by companies as a software-as-a-service (SaaS) product. Therefore, we have a digital twin-as-a-service (DTaaS) that can be designed and sold in an enterprise model.

2 Digital Twin Benefits and Challenges for the Industry

DT is still in its developing phase, and challenges are inevitable, but it also brings numerous opportunities for industrial applications. The implementation of DT leads to new discoveries, more accurate results, and significant improvements in efficiency, sustainability, reliability, and long-term cost reduction. DT opportunities can be categorized as follows [1, 19]:

1. Real-time monitoring: By incorporating more input data into the analytical toolbox, DT provides a comprehensive and detailed understanding of the physical object, utilizing various visualization techniques to enhance user visibility. DT offers continuous remote monitoring of the object, asset, system, or process, enabling the prediction of forthcoming issues (e.g., component aging). This proactive approach significantly minimizes downtime and reduces maintenance costs.
2. Optimal operation: The computational analysis of DT considers all relevant information about the object's condition and behavior, enabling it to derive the optimal operation. This approach identifies potential areas for improvement, resulting in enhanced efficiency and safety.
3. Improve quality: DT analysis can uncover hidden potential that enables efficient design and development. Three-dimensional models of DT deliver a wide range of aspects related to the product life cycle. The updated settings of the real object and its related phenomena directly impact quality.
4. Innovation: DT enables innovation in various applications. Firstly, it can be used to design new and enhanced products. Secondly, DT can simulate new approaches to recreate production steps. Additionally, suggested supporting solutions of DT help engineers discover and test new experiments.
5. Strong maintenance solutions: The predictive analysis feature of DT provides a near-future picture of the product and a long-term perspective of its behavior. Data analysis, especially when involving multiple data sources, enhances self-awareness of the machine. The predictive analysis, combined with virtual troubleshooting, offers maintenance and reconstruction solutions and a continuously updating schedule to ensure the asset's optimal functionality throughout its lifespan.

6. Reducing energy consumption: DT improves automation in several industry applications, inherently planning, organizing, controlling, and modifying the entire operation to save energy consumption. DT can be specially designed to manage and control energy consumption in buildings.
7. Cost reduction: DT offers operational optimization, reducing operation costs, predictive maintenance, and fault identification, which reduce breakdown and repair costs. Additionally, DT enhances workforce productivity, balancing company expenses.
8. Reducing time to market: DT enriches manufacturing companies with efficient product design, the ability to assess an item before production, understanding and preventing possible failures, and risk analysis and management. These factors result in reduced time to market.
9. Increasing user engagement: A DT platform should be accessible from anywhere and provide intellectual interaction with users, essential for safety, especially in high-risk environments. Digital devices, such as smartphones, tablets, and headsets, are common devices to install DT on and assist users. Moreover, DT simplifies personalization of products and services in manufacturing lines.

Including pioneering and high-influence technologies, DT shares challenges in parallel with its building modules. The articulated shortcomings and barriers in the design and implementation of a DT are as follows:

1. Immature building technologies: Most of the technologies in DT architecture are still under development and not fully mature. For instance, the lack of powerful computing platforms, limited IoT device battery lifespans, AI complications, cyberattacks, and security breaches, as well as uncompleted business models, hinder the progress. Furthermore, the absence of ubiquitous and high-speed communication networks exacerbates the limitations of aging legacy systems, challenges in infrastructure changes, and high costs.
2. Data issues [20]: Data is harvested from different resources; hence, heterogeneous and compatible data types and frameworks are the bottleneck of many data-centric technologies. Other data-related complications include inefficient data governance methods, trust, privacy, cybersecurity, convergence, acquisition, large-scale analysis, modeling, and big data issues. Moreover, modeling real-time and bidirectional information flows is particularly challenging in complex DT structures.
3. Incomplete standards: Implementing a standard DT requires international technical standard specifications that describe connections, models, interfaces, protocols, and platforms. Researchers in academia and industry should cooperatively develop and expedite acceptable standards and regulations.
4. Infrastructures: Various infrastructure challenges arise due to different IT manufacturers providing a variety of commercialized instruments and devices. This leads to hurdles of integration, scalability, and interoperability. Human interface devices, connectivity, internet services, sensors, and their battery lifespan are examples of undeveloped configurations. Moreover, costs and budget constraints

in both hardware and software infrastructures make DT implementation expensive and inaccessible.
5. Lack of digital skills: Insufficient knowledge of DT architecture and building technologies among engineers, designers, and common users hinders successful implementation of DT. Accordingly, a proper insight into complex DT technology requires investment in training programs to upskill more expertise.

Chapter 3
Digital Twin: A Novel Solution for Smart Grid Applications

The usual and known electricity network contains large power plants that generate energy in remote areas, long transmission lines to load centers in cities or industrial areas, and power distribution systems that receive and distribute it to load points.

Conventional power systems have evolved alongside technological advancements. Much of this development has been achieved through improving current mechanisms, such as developing power markets to create competitive environments for power plants and energy consumers.

Smart grid affects all sections of the power system, as shown in Fig. 3.1, accommodating structural and operational alters. For example, renewable resources owned by consumers can generate power locally, and excess energy can be injected into the power system and used in other load points.

Migrating from traditional power systems to smart grids necessitates a comprehensive consideration of all existing and emerging domains. As outlined in the IEEE 2030 standard [8], smart grid fields are categorized into seven distinct domains: generation, transmission, distribution, customer, operations, service provider, and markets. Despite significant advancements in these domains, including the development of appliances and the installation of new equipment such as PMUs and FACTS, there remains a substantial gap to be bridged before achieving a satisfactory level of performance.

Since smart grid (SG) enhances the traditional power grid by integrating intelligence and Information and Communication Technology (ICT) components, digital twin (DT) can digitally replicate and optimize the entire system, including its functionalities, processes, and services. DT can be designed for the following significant purposes in SG:

- Inspection, monitoring, and maintenance.
- Energy management and monitoring.
- Precise load control/demand response.
- AMI, DERs, vehicle communications.

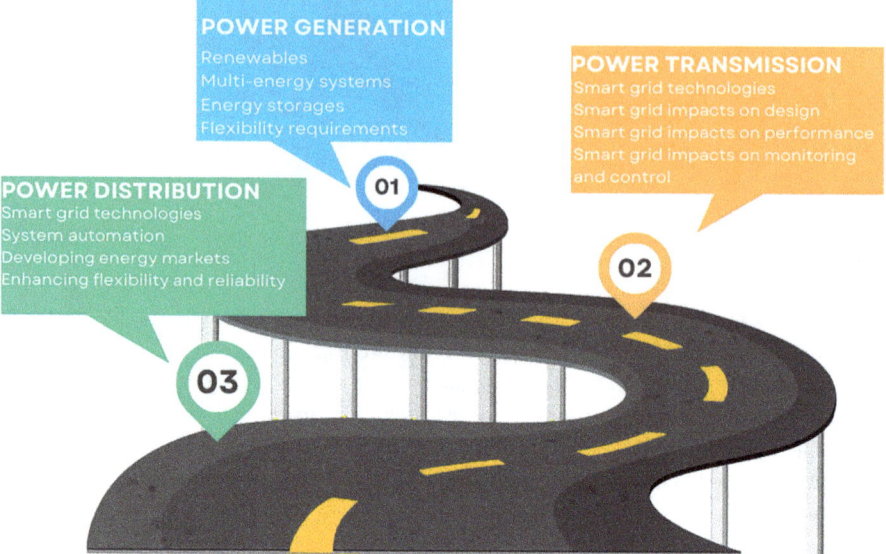

Fig. 3.1 The effect of smart grid concept on different sections of the power system

- Fast data transmission for video monitoring.
- Security-sensitive and protection applications.
- Usage management.
- Real-time price of electricity broadcasting.

1 Smart Grid Structure

SG has introduced innovative strategies and leveraged cutting-edge technologies to enhance traditional energy generation, transmission, and distribution processes. Figure 3.2 presents a comprehensive overview of SG configuration, providing a visual representation of SG structure.

The above structure indicates that some integrated generating units (VPP), microgrids (MGs), and storage units could connect to SG. These concepts enable the SG to incorporate different types of distributed generation resources. Small-scale photovoltaic panels will be installed on rooftops and exchange energy with low-voltage levels of SG. Smart measuring units and electric cars are shown in residential areas as well. Wholesale and local electricity markets are structures that facilitate economic exchanges of power between different entities.

1 Smart Grid Structure

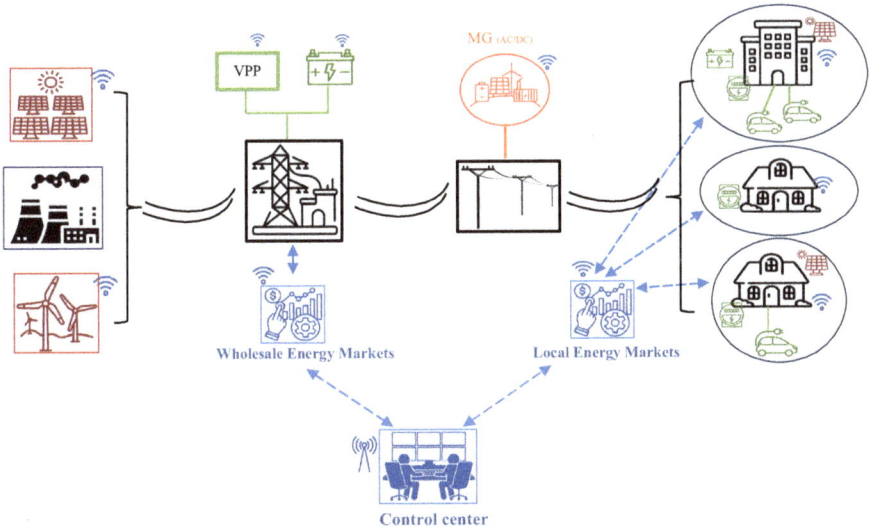

Fig. 3.2 Smart grid structure

One key difference between SG and legacy power systems is the role of data in system operation procedures. The most commonly used icons in the SG figure are communication blocks that facilitate data transfer, including generation level of plants, energy usage in different areas, market price, control and monitoring information, weather conditions, and security data. This shows significant effect and heavy duty of ICT and digital technologies in advanced power systems.

Another fresh context in SG is a new role called prosumer. High penetration of renewable energy resources in power distribution systems such as rooftop solar panels introduces prosumer (i.e., producer + consumer). At specific times, a prosumer may supply all of its load and also sell excess energy to the power distribution grid. In other words, some prosumers sometimes act as power resources when the energy generation of their installed equipment is higher than their energy demand.

Note that communication infrastructures that are primarily used in SG are as follows:

- Cable networks: Optical fiber technology enables high-bandwidth, low-latency, and secure data transmission via non-metallic fiber cables, utilizing pulses of light signals. Many electric utilities have already installed and utilized dedicated fiber cables. Additionally, power line communication (PLC) leverages existing power lines as a data transmission medium, offering low costs, scalability, and stability. Both cable networks provide reliable and secure infrastructures for critical management issues.
- Wireless networks have revolutionized communication by eliminating the need for physical wires or cables. Instead, they utilize radio waves to facilitate low-latency data transfer and promote mobility, flexibility, and scalability. This technology includes mobile cellular networks, radio broadcasting, and WLAN or

Wi-Fi networks. These technologies play crucial roles in modern SG, such as gathering metering information and connecting electric vehicle (EV) charging stations.

2 Digitalization of Smart Grid Subsystems

The cyber-physical nature of SG arises from integration of electric networks and ICT techniques. Given the numerous mission-critical processes within the SG ecosystem, digitalization has significant potential to optimize operations and accelerate development in SG applications.

Digitalization technologies that enhanced SG subsystems have prompted procedures in two groups, as discussed below.

2.1 Sensing and Measurement

The data sensing, transmission, and collection features of the power network provide real-time or near-real-time high-precision data on the physical state and condition of network components, including voltage, current, frequency, power flow, and temperature. Advanced sensing and measurement technologies enable data analysis and intelligence, which facilitates fault detection, optimization, prediction, and restoration. Examples of sensing and measurement networks in SG are as follows:

- Phasor measurement units (PMUs): A network of PMUs installed on various points of power lines measure voltage and current phasors with high accuracy and synchronization. They provide situational awareness, state estimation, stability analysis, wide-area monitoring, protection, and control. Synchro-phasor networks collect, transmit, process, supervise, store, control, and display PMU data. They allow real-time analysis at high speeds, faster than current control systems, and provide a comprehensive view of power system dynamics.
- Smart meters: New generations of meters measure and record electricity consumption or generation of customers or distributed energy resources (DERs) at different intervals, providing demand response signals, load profiling, outage detection, billing information, and more.

2.2 Control and Automation

Intelligent control actions optimize and regulate the operation of components or subsystems within SG. Moreover, automation tries to reduce human intervention. Both control and automation approaches also facilitate self-healing capabilities and adaptive protection schemes.

In the energy sector, flexible AC transmission systems (FACTS) are a prime example of control and automation technologies that enable rapid and flexible control of various parameters in transmission networks, including voltage, impedance, and power flow. FACTS devices incorporate power electronic components that enhance the stability, security, efficiency, and power transfer capability of power transmission networks by compensating for reactive power, regulating voltage profiles, controlling power flow, damping oscillations, and more. Furthermore, FACTS devices facilitate the integration of distributed energy resources into power transmission networks by providing benefits such as voltage support, frequency control, and power quality improvement.

Digital energy transformation in SG occurs across the energy value chain and in all aspects of the energy system. The numerous benefits of digitalization in SG have been extensively proved, and lastly, DT technology offers enhancements across all SG domains. In the next three sections, we will describe new structures of power generation, transmission, and distribution sectors in SG, and explain how the novel DT technology can provide advanced benefits in every SG subsystem.

Chapter 4
Digital Twin for Smart Power Generation

According to [21], annual electrical power generation in the world will increase to 42,500 TWh by 2040. It has been about 23,800 TWh in the year 2015. SG integrates renewable energy resources into modern power networks to enhance overall power generation and improve the sustainability and resilience of the energy system. Leveraging communication networks, sensors, and data analysis, SG upgrades real-time control and monitoring of power generation resources, thereby increasing performance, security, and reliability across the entire power system.

This section focuses on the impact of DT technology on plants and energy production. DT offers significant benefits to the generation sector, delivering threefold values. Firstly, DT enhances energy production by predicting and detecting equipment failures, thereby ensuring uninterrupted operation. Secondly, DT increases the efficiency and sustainability of generation units by optimizing their performance. Thirdly, DT establishes proper operation and maintenance (O&M) plans, leveraging AI to suggest preventive O&M schemes for various power plants.

Hereunder, we study four primary areas of electricity generation (Fig. 4.1) and discuss the benefits that DT offers to them.

1 Conventional Power Plants

Traditional power generation technologies involve large-scale thermal power plants that consume fossil fuels, uranium, and coal. These plants contain a complex system with several intricate and expensive equipment that have been used over time.

Operation and maintenance of these large structures have been prioritized to optimize their productivity, minimize downtime, and reduce servicing costs. Recently, monitoring systems have been designed and installed in power plants using central control smart devices. These structures will continue to play a crucial

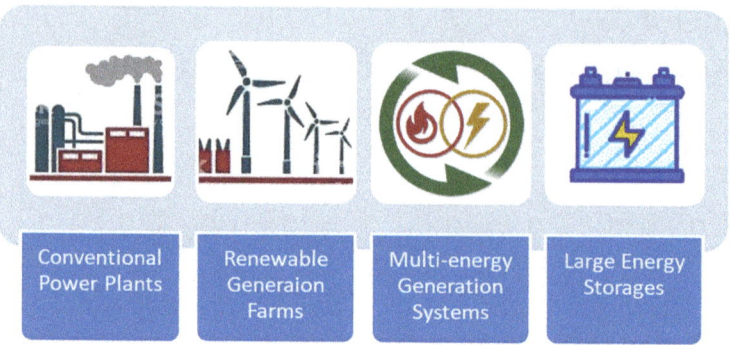

Fig. 4.1 Smart power generation subsystems

role in the SG environment for years to come, as they remain the primary sources of electricity in many countries.

Numerous industrial and technological companies, research centers, and academic institutions have extensively studied DT technology to continuously monitor and analyze plant activities, identifying trends that predict potential issues. Notable examples of successful DT implementations include general electric (GE) and Siemens, which have developed DT solutions for key modules and properties such as turbines, boilers, and compressors.

DT can effectively assess performance level of power plants, particularly when leveraging AI and machine learning (ML) techniques. By employing DT, plants can benefit from efficient methods of functionality analysis, identifying defective components, and scheduling predictive maintenance with minimal production disruptions. Furthermore, DT also suggests improvements to existing operations and enhances the efficiency of power generation facilities.

2 Renewable Power Generation

Fossil fuels, the primary energy source for thermal power plants, are not universally available, and governments must often rely on international energy markets to secure their supply. So, conventional power plants are gradually replaced by renewables, especially solar and wind technologies. Renewable power plants use primary energy resources that can be regenerated in nature. Some prominent examples are hydro, solar, and wind power generation technologies. Unlike traditional energy supplies, one distinctive feature of renewable power resources is their availability in most areas of the world [22, 23].

Integration of legacy generation systems with lots of renewable units, alongside modern ICT modules that interface with established control and management systems, poses significant challenges. Currently, DT can adjust these renewables within SG, which are alternative power generation solutions.

Generally, DT developing strategies benefit renewable generation units by increasing robustness, facilitating interoperability of power generation units, indicating root causes of faults or losses, predetermining failures or malfunctions, managing assets, inspection and monitoring, and cutting maintenance costs.

2.1 Solar Panels

Solar power generation technologies utilize sunlight energy to produce electrical power through two approaches: photovoltaic (PV) panels and concentrated solar power (CSP). In PV technology, solar panels are composed of solar cells that directly convert sunlight into electricity using semiconductor materials. Inverters and batteries play a crucial role in PV systems, enabling the storage of generated energy or meeting energy demands. The angle of solar radiation on solar panels significantly impacts the output power of photovoltaic systems. Compared to other power generation technologies, PV systems have lower maintenance costs due to the absence of rotational parts. Integration of PV systems with energy storage systems (to store excess energy of PV systems) enhances their performance, particularly in power grids with peak demand during nighttime hours.

CSP is another type of solar power generation that employs mirrors to concentrate sunlight onto a focal point, heating a fluid within a tank or tube. This heated fluid is then utilized in other cycles to power a thermal power plant. Moreover, it can be stored in thermal storage systems for later use during the day or night, ensuring a constant supply of energy. As a result, CSP technology is available at all times.

However, PV systems are more prevalent globally due to their widespread installation. PV systems are simpler and more versatile than CSP, allowing for smaller-scale installations such as rooftop systems. In contrast, CSP systems require a significant area for the installation of mirrors and lenses. Additionally, PV systems have lower maintenance costs and can generate power in low-light conditions.

As the solar industry experiences rapid growth globally, it is driven by the development of advanced technologies, including DT. A solar system is susceptible to various suppressions, necessitating a robust supervisory unit like DT that can monitor the overall performance of PV panels and the functionality of individual components and panels. As depicted in Fig. 4.2, design, performance monitoring, lifecycle management, maintenance scheduling, safety, and security are the main profits of DT for solar sectors.

Initially, a DT can be introduced to prosumers and rooftop panel owners through a software application on personal computers or mobile devices. This DT system can provide various essential information about panel's functionality, optimal maintenance schedules, and future energy generation forecasts. By engaging customers and offering various information, DT ensures customers that panels operate at their best and they can effectively manage electricity generation.

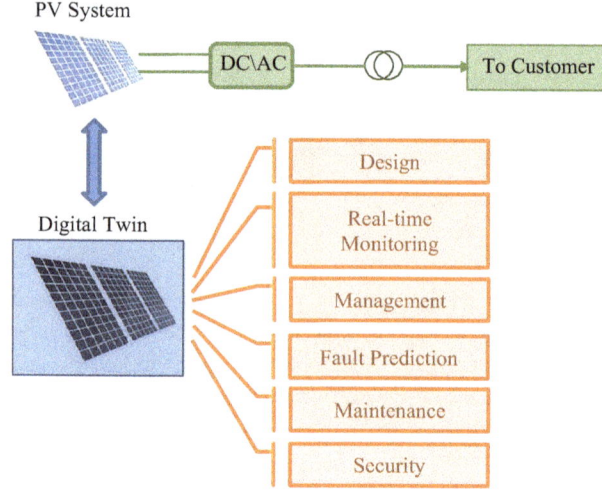

Fig. 4.2 DT advantages to PV systems

2.2 Wind Turbines

Wind power plants, comprising turbines with distinct categories based on construction, generator type, grid connection, and control systems, have evolved significantly. Notably, the most powerful commercially deployed wind turbine, the GE Haiade-X [24], boasts an impressive 11 MW generation capacity, making it a prominent offshore turbine by 2022.

Early modern wind turbines had a constant speed regardless of wind speed. Their generators were directly connected to the grid. Although they had a simple structure and low prices, their main disadvantages were the inability to control voltage, lower efficiency in non-rated wind speeds, and the transmission of noise to the power network in case of wind speed variations. However, current vendors produce variable-speed wind turbines using a pitch control mechanism, in which the angle of the turbine's blades is adjusted to reach an appropriate speed and generate more electrical power up to the rated speed.

Considering variable rotor speed of wind turbines, two solutions are available: doubly fed induction generators (DFIGs) and frequency converters coupled with synchronous generators. It is important to note that frequency converters employ power electronic technologies, specifically insulated gate bipolar transistors (IGBTs), as mentioned in [25].

In general, there are two series of converters, one on the rotor side and one on the grid side of turbines, used for several purposes such as extracting maximum power from the turbine by controlling the rotor speed, and also controlling the voltage of the connecting bus of the grid by generating reactive power.

Recent advancements in wind turbine technology have enabled the integration of active and reactive power control strategies, allowing these power plants to participate in voltage and frequency control structures in future power systems with high levels of renewable energy penetration.

Installing large-scale wind power plants with significant capacity requires providing backup capacity to increase the flexibility of the system. This is necessary to recover from instant fluctuations in their power generation, which may lead to additional costs.

A wind power station consists of multiple wind turbines, which necessitate individual monitoring and operation, as well as overall plant management. Wind turbines are massive structures typically installed in challenging geographical conditions. The requirement for monitoring each individual turbine and its components, as well as the entire farm operation, prompts engineers to develop software systems that enable remote performance monitoring.

Wind turbines are one of the first focused elements for developing a DT platform. GE, a significant energy technology company, was a pioneer in designing DT at the industry level and initiated a DT for wind turbines.

With several innovative features of a DT, it actuates various advantages for wind turbines based on sensor measurements. Some of these benefits are general, and others can be classified based on components. The former encompasses efficient data processing, more precise model updates, novel feature extraction, accelerated performance analysis, real-time state estimation, and condition monitoring systems. Additionally, it includes cost-effective maintenance strategies, such as predictive and preventive schedules, which reduce virtual test time and extend the remaining useful life of the wind turbine. Figure 4.3 illustrates the benefits of DT for the constituent parts of a wind turbine.

3 Multi-energy Generation Systems

Two prominent types of distributed generation technologies in multi-energy generation systems are combined heat and power (CHP) and combined cooling, heating, and power (CCHP). These co-generation systems offer substantial benefits in terms of energy efficiency and environmental performance. CHP produces electricity and useful heat simultaneously from a single fuel source, whereas CCHP generates cooling, in addition to heating and power. These systems can achieve high overall efficiencies, resulting in substantial reductions in greenhouse gas emissions and energy costs.

In comparison to traditional single-source systems, multi-energy generation systems, particularly CHP, offer several advantages, which are summarized below:

- They are highly efficient, as they utilize waste heat from the power generation process and supply thermal loads.
- They enhance the reliability and resiliency of the energy system by allowing for seamless transitions between different energy resources in the event of system contingencies.

Fig. 4.3 DT solutions for wind turbines

- They minimize emissions by leveraging renewable energy sources or optimizing fuel consumption for thermal applications.
- They improve economic efficiency by using the price difference of energy resources during operation horizon.
- They increase system stability by providing additional capacity for the power system, especially during peak demand periods.
- They expand in remote areas to supply demands in off-grid mode.
- They quickly adapt to changes in demand by leveraging the performance of various energy resources, thereby enhancing the system's flexibility.

Recent research has investigated the integration of CHP systems with renewable energy sources [26], including biomass-based CHP [27], geothermal-based CHP [28], and hydrogen-based CHP. Moreover, some energy generation systems use organic matter as fuel, such as waste-to-energy (WTE) [29] and biofuel plants [30, 31].

Geothermal power plants are another renewable energy resource that harnesses the natural heat from the Earth's crust to generate electricity [32]. These plants are typically located in areas with high geothermal activity. To operate, geothermal power plants require drilling into the Earth's crust to access hot water and steam, which are then used to drive a turbine and produce electricity. From an environmental perspective, geothermal power plants are considered clean technologies, emitting very little greenhouse gas. There are several types of geothermal technologies, including dry steam, flash steam, and binary cycle power plants.

In general, DT solutions for multi-energy generation systems, similar to those discussed above, follow optimization principles that are applicable to these structures. Furthermore, the smart infrastructure platform of DT performs performance analysis to investigate and suggest optimal operating conditions, thereby optimizing the functionalities of energy assets. Additionally, DT follows energy market volatility to develop the best strategy for multi-energy generation units, aiming to minimize costs as much as possible.

4 Energy Storages

Energy storage in SG plays a crucial role in providing ancillary services to electricity networks and supporting their growth. Scalable storage units enhance the flexibility of SG by shifting the timing of supply, allowing for more efficient management of energy distribution. Grid-scale batteries have experienced significant growth in recent years, offering a reliable and efficient means of energy storage. Computer-based technologies provide solutions for monitoring and regulating the flow of electricity from various sources of generation to storage units, enabling the efficient meeting of varying consumer demands.

Various energy storage systems can be utilized in intelligent power systems. For instance, Hydrogen is a developing technology with potential for seasonal renewable energy storage. Leading storage technologies are glanced as follows:

(a) Pumped-Storage Hydropower and Gravity Storage: Pumped-storage hydropower (PSH) is an efficient method of energy storage that leverages the principles of water and gravity. This is the most widespread storage technology and has significant untapped potential in many areas. PSH has two water reservoirs, and water is pumped from the lower reservoir to the upper one in case of excess or cheap electricity. Subsequently, the energy is released through turbines to generate electricity, particularly during periods of high electricity prices or peak demand [33]. PSH enhances power system flexibility, allowing system operators to utilize more capacity from intermittent renewable power generation technologies, such as wind and solar. In a broader sense, gravity storage can utilize various objects or water to store energy, such as water in two tanks at different heights, concrete blocks on a tower, weights on a cable, or water on a mountain.

(b) Compressed air energy storage (CAES): Electrical energy can be stored as compressed air in CAES units. CAES consumes electricity to compress air and store it in underground reservoirs or overground tanks [34]. When needed, the compressed air is released through a turbine and generates electricity. A critical advantage of CAES is its capability to store large amounts of electricity for several days or weeks. Additionally, these systems can be installed in various scales, making them suitable for a range of applications, from small residential to large utility systems. Moreover, the investment and operation costs of CAES

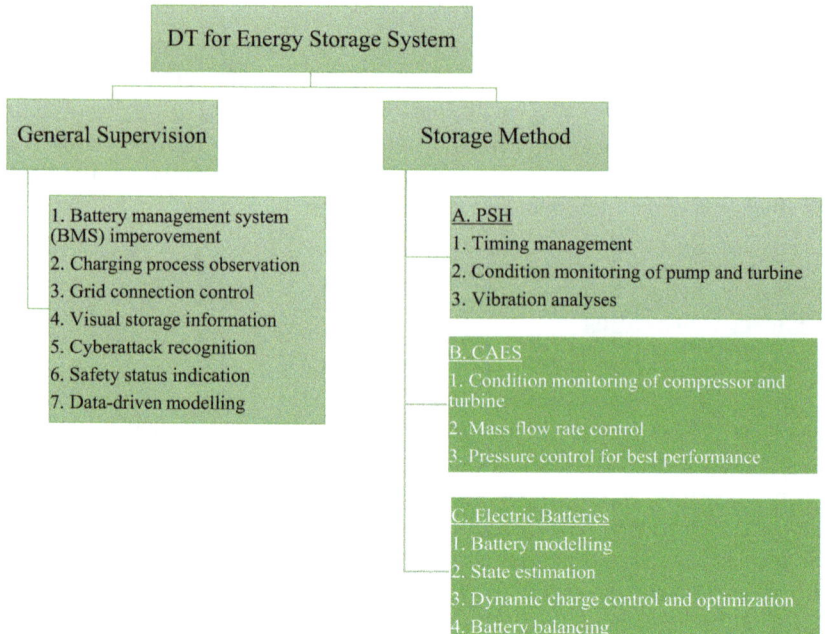

Fig. 4.4 DT opportunities for energy storage systems

are relatively low. In environmental terms, CAES is an eco-friendly technology, as it enhances system flexibility, enabling the integration of more renewable resources and reducing reliance on fossil fuels [35].

(c) Electric batteries: These storage units store and generate energy through various chemical reactions. Due to the increasing adoption of electric vehicles (EVs), which represent a new type of electrical energy consumer, there has been a greater focus on electric batteries.

Electrical batteries come in various types, including lead-acid, nickel-cadmium (NiCd), nickel-metal hydride (NiMH), lithium-ion (Li-ion), and lithium-polymer (LiPo). Among these, lead-acid batteries are the oldest and most commonly used rechargeable batteries, utilizing lead plates as electrodes and sulfuric acid as the electrolyte. They are widely used in vehicles, backup power systems, and industrial applications. Their main advantages are that they are cost-effective and powerful.

Primarily, DT provides intelligence to battery management systems [19] by offering recommendations based on various information and parameters. As illustrated in Fig. 4.4, we categorize DT opportunities for energy storage systems into two groups: general supervision, which benefits any type of energy storage technology, and specific supervision, which is tailored to each storage technique.

Chapter 5
Digital Twin for Smart Power Distribution

Power distribution system is responsible for delivering energy to end-users. In the context of SG, analyzing the functions of power distribution system that directly supply consumers is of great importance. Smart distribution network should provide suitable conditions for energy consumers to cooperate effectively with distribution system operators (DSOs).

Smart power distribution system presents both challenges and opportunities. The primary challenge lies in the integration of renewable power generation units into the existing distribution system. This poses significant challenges for DSOs as these resources are highly intermittent and require appropriate monitoring and control instruments. Cybersecurity is another significant challenge because it necessitates two-way communication links, making confidential information susceptible to breaches and theft. Cyberattacks can disrupt the system and reduce its security. Additionally, upgrading power distribution infrastructure to meet SG standards will be costly due to aging devices. Finally, consumer acceptance is another challenge as not all consumers are aware of benefits of SG applications such as load management, and may not be willing to change their behavior to participate in programs.

SG concept offers valuable opportunities for power distribution systems. Real-time monitoring and control defined by SG improve reliability and efficiency. Distributed energy resources such as rooftop solar panels, energy storage units, and demand response programs increases energy efficiency by reducing power loss during transmission from large power plants to load centers. Data gathering and analysis are other essential demands of SG. Analyzing power generation and consumption data provides functional patterns for DSOs or planners to make appropriate decisions, such as designing demand response programs based on consumers' energy price sensitivities.

Since DSO is the main controller of distribution system in SG, DT will assist its management responsibilities by providing a clear and comprehensive view of subsystems. DT, authorized by distribution data centers, will establish conjunction and collaboration of distribution units (e.g., prosumers, microgrids, and energy markets)

Fig. 5.1 Smart power distribution subsystems

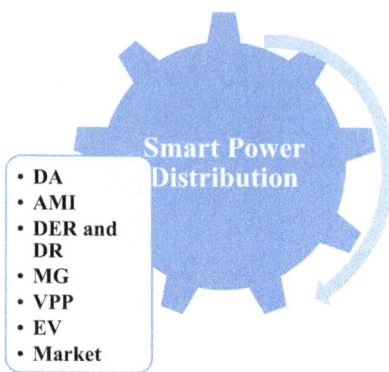

without the need for human interventions. This will enable the DT to perform condition evaluation, fault prediction, and asset management duties.

Primary distribution subsystems detailed in SG, are illustrated in Fig. 5.1.

1 Distribution Automation (DA)

DA brings intelligence to power distribution system in SG, mostly in subsystems and utilities. In smart power distribution systems, DA leverages hardware devices, primarily sensors and communication equipment, along with software tools to collect and analyze data. This enables DA to perform automated functions that monitor and regulate the electricity distribution network.

Enhanced operational efficiency and reliability of the power distribution system, rapid identification of exact fault location, and reduced load interruption times are among primary benefits of DA.

Some of the main functions of DA in SG are as following:

- Voltage control of buses.
- Power flow control of feeders.
- Remote control of system topology.
- Automatic detection of fault location and power restoration.

DA includes a wide range of activities, for advanced monitoring and control actions. For example, advanced fault location, isolation, and service restoration (FLISR) technique is self-healing process of DA designed to detect and respond to any outages. In this process, location of system faults is determined, and then automatic control actions are performed on switches to isolate the faulted sections and restore interrupted loads optimally. This improves reliability and resiliency of distribution system.

In addition to technologies like FLISR that respond to critical system faults, DA has many applications related to normal state system operation. DA monitors voltage levels of buses, which fluctuate due to load variations. It then automatically switches capacitor banks to control voltage levels within the allowed range and reconfigures automatic feeders as needed.

DA has converged to digital frameworks and is now being empowered by DT support and profits. Data-oriented utility automation already contains digital techniques and has the potential to conduct impact analysis. At the same time, DT relies on high-precision models and simulates smart distribution systems in real time. Therefore, DT can expand DA to use a more comprehensive database, utilize learning methods, draw out generalized outcomes, visualize results, and optimize functional decisions. DT can cooperate with FLISR modules to suggest more accurate fault detection mechanisms and prevent and handle probable outages.

2 Advanced Metering Infrastructure (AMI)

AMI is part of automation penetration in energy distribution systems. It is an advanced technology that enables utilities to collect energy information and prosumers' data. Unlike traditional monitoring systems, AMI provides real-time monitoring that increases reliability and efficiency. In addition to utilities, consumers can also use AMI data to be informed of their energy uses and make appropriate decisions.

Smart meters are fundamental components of AMI that collect various immediate data related to energy consumption and transmit them to the utility in real time. Some of the valuable data metered by smart meters include voltage profile, current, active and reactive power, and power factor. This information is then sent through a communication network to the utility for further analysis and optimization. AMI employs a two-way communication network for consumers and utility interconnection. After receiving and analyzing data of smart meters, AMI sends back some controlling or warning signals to end-users to inform them about their energy usage or commands to connect/disconnect their energy systems.

Some countries use real-time pricing of electricity in which electricity prices are calculated in short-term horizons of 5 or 15 min and power producers and consumers are cleared based on these prices. AMI can send real-time electricity prices to data panel of consumers. AMI can send real-time electricity pricing information to the data panels or displays of consumers in their homes or businesses. So, they can manage energy usage profiles to reduce their energy cost. In addition, AMI plays a crucial role in implementing demand-side management (DSM) programs, such as emergency demand response programs (EDRP). In an EDRP, customers enter into a contract with the distribution company, granting it the authority to temporarily disconnect and reconnect specific equipment during periods of high electricity consumption or emergency situations.

AMI appliances help prosumers participate in local energy markets and sell their excess energy or buy energy if needed. Concepts such as peer-to-peer (P2P) trading are defined based on prosumers' performance in which they directly trade energy.

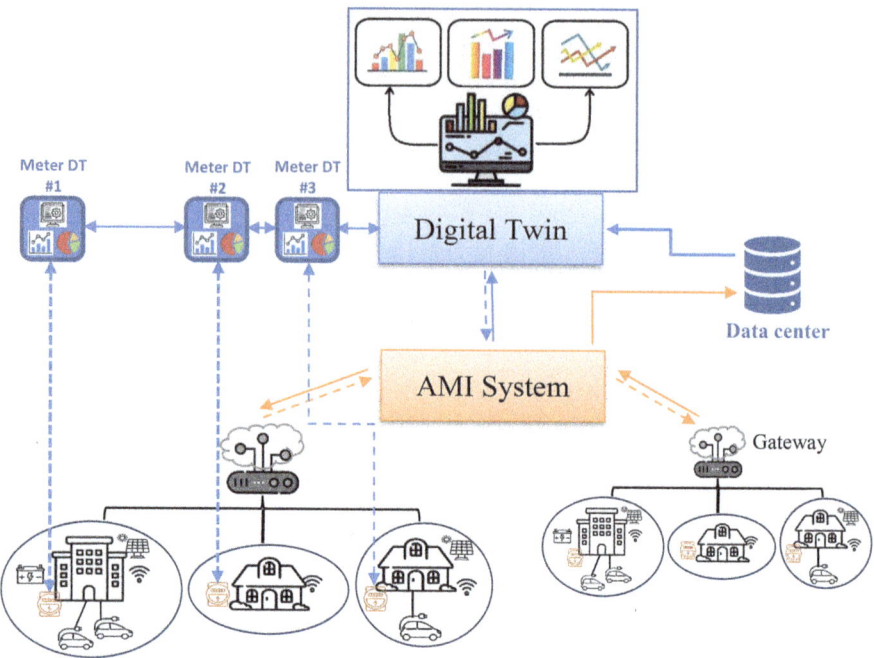

Fig. 5.2 DT and AMI cooperative system

Such transactions carried out by smart homes can suddenly reduce energy costs because they are done without intermediaries and between final consumers of energy.

AMI can be seen as a foundation for DT, which collects data and transmits it to an aggregation center through gateways (Fig. 5.2). Here, a DT operates based on smart meter data within a local energy system. Conversely, DT can be applied to enhance the AMI system primarily in AMI data analysis and autonomy, allowing individual meters to perform self-governance and enhancing AMI interaction with local management systems (e.g., DSO), if necessary. As a result, DT technology drives the development of the AMI system, and the benefits are mutually advantageous.

As illustrated in Fig. 5.2, DTaaS can be developed in modules for individual customers to monitor their energy consumption and receive operational information advice. Interconnection and interoperability of DT modules at various scales, whether in a P2P fashion or across hierarchical multiple levels, establish a networked DT system. This networked-DT approach offers several noteworthy advantages such as more accurate load profiles, real-time information of feeders to feed DA functions, forecasting, and estimation.

3 Distributed Energy Resources (DERs) and Demand Response (DR) Programs

Distributed energy resources (DERs) are small-scale power generation units connected to the power distribution system. These units include renewable energy sources, energy storages, and electric vehicles. Unlike centralized architectures, diversified DER units are integrated into the distribution network to foster a decentralized SG structure.

In addition, demand response (DR) is a type of DER that enables consumers to actively participate in system operation procedures and would allow utilities to manage power demand in peak periods [36].

DR programs include various services. The first is time-of-use (TOU) pricing, which determines energy prices for peak and off-peak periods. Consumers can control their consumption time to reduce energy costs. The second is critical peak pricing (CPP), which is similar to TOU in pricing, but with a more dynamic pricing strategy based on load levels. Another DR program is direct load control (DLC), where system operators can remotely control and reduce consumers' loads during peak times. Participating consumers are eligible to receive incentives in return. Next, there are interruptible load programs (ILPs), where consumers sign a contract with the system operator to cut off their energy consumption during periods of peak demand. Finally, there is the emergency demand response program (EDRP), as described in the AMI subsection, which is activated when the system operator announces an emerging request for load reduction.

Due to the inherent variability and uncertainty of DERs, their output power changes rapidly and unpredictably, often based on the uncertainties associated with weather conditions. This behavior creates a mismatch between the power supply and electricity demand, which can subsequently reduce power quality, reliability, and security. To address these challenges, the power system should have sufficient flexible capacity on both supply side and demand side.

Individual DT platforms tailored for any of DER modules initiate significant benefits to DER holders, such as observability, scalability, and cost management. The point is that several small-scale DTs can come together at the distribution level. As described for AMI, if multiple DTs that service separate prosumers collaborate, they send comprehensive regional information toward DSO. This information will serve as a foundation for analytical studies. Here, the main advantage becomes evident since utilities can observe the entire area, monitor their timely production and market involvement of prosumers, and then make best decisions at management level. For example, they can better balance power plants' production for the grid.

The networked-DT of DERs has a highly dynamic structure, which brings benefits to both individual DER units and the overall distribution grid. It increases sustainability, helps physical and financial risk management, and develops energy resiliency. DER is a broader term than DR, hence, DT opportunities for DER can be discussed for DR programs as well. Mutual DR advantages for both electricity prosumers and grid operators improve when stronger and more diverse data analysis results of DT are available and prepared.

4 Microgrids (MG)

MGs are local, small-scale power distribution systems within the SG that integrate various types of DERs, storage units, and loads. These MGs can operate independently from the main grid, supplying their local loads. To ensure optimal operation, MGs are typically equipped with advanced monitoring and control tools. These tools enable MG to operate at its optimal working point, considering both normal operating conditions and contingency scenarios. In the event of a fault, MG can be isolated from the main grid to maintain reliable and uninterrupted power supply to its local loads. They usually have an energy management system (EMS) to determine the best decisions in different operation states [37].

Compared to conventional power distribution systems, MGs increase reliability, flexibility, and security based on available resources. As MGs typically have a lower number of load points, they can model load properties in greater detail, such as load vulnerability models against power interruption [38–41]. In addition to electrical loads, MGs may also include thermal loads that need to be supplied by CHP units and boilers. Modern microgrids empower system operators to address data uncertainties, such as forecasts of load and renewable power generation, by employing advanced monitoring tools, data analysis techniques, and decision-making algorithms.

Power distribution systems can be regarded as a community of connected MGs. Each MG is an autonomous power system supplying its internal demand and can exchange energy with distribution system. In normal operation state, MGs are connected to the main grid and aim to operate in an economically efficient manner. In case of main grid faults, MGs get disconnected and supply local demand in island mode. Developing multi-energy MGs can enhance system reliability and flexibility, as they can rapidly replace other types of energy resources in the event of failures. For instance, if disconnected from the main grid, electrical demand can be promptly supplied by DGs or CHPs. Various energy-related, power-related, and ramping-related flexibility metrics can also be improved by MGs [42, 43].

As expected, shown in Fig. 5.3, DT can improve MG operation and support their prominent functionalities. Let's extend the organization of DT opportunities for MGs pursuing description in [44]:

1. Advanced real-time monitoring: To monitor and supervise regional MG subsystems and items, DT upgrades MG's head management system and introduces MG as a controllable energy entity. DT maximizes DERs' capacity, construct a well-balanced energy network and improves its flexibility and reliability.
2. Security and Protection: Various anomalies might occur at different operational levels. DT provides a compliant protection scheme for any structure of MG. Thus, control of MG units, fault recognition, and protection plans are presented in DT visual schemes. Furthermore, security protocols are integrated into DT models to maintain a trustworthy and stable model of established MG.
3. Predictive maintenance: To keep assets running efficiently, systematic maintenance policies preserve and optimize facilities' functionality. It'd be best if maintenance service predicts potential defects and predetermines possible fail-

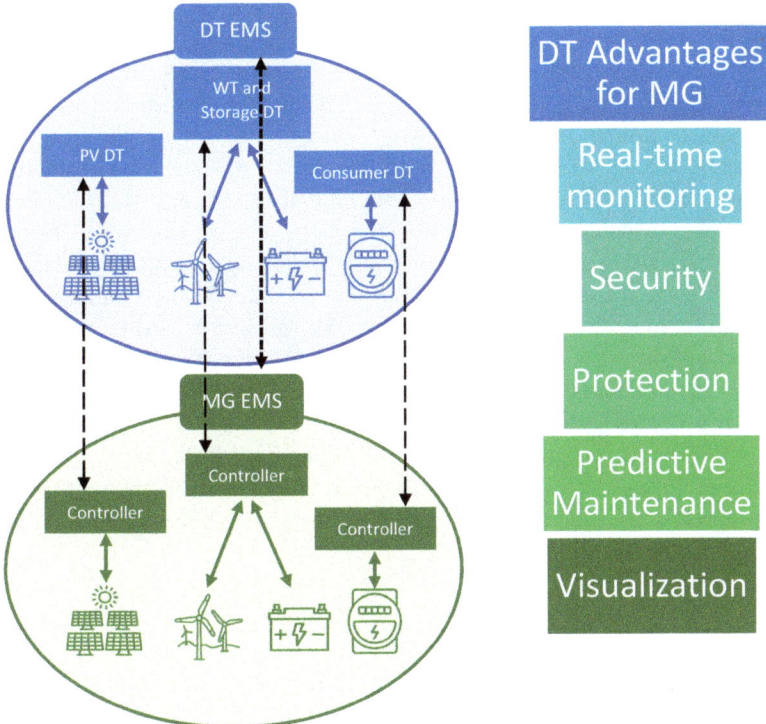

Fig. 5.3 DT image of MG and its opportunities

ures. DTs possess the capability to monitor and analyze machine downtimes, forecast and notify potential deteriorations, and recommend timely preventive steps to schedule maintenance activities for extending lifespan of machineries and mitigate future operational risks, ultimately leading to reduced lifecycle costs.

4. Visualized simulation Results: DT generates images from each partial component, and an extensive snapshot of the entire MG. These visualized results are then utilized for behavioral analysis, functional evaluation, and adaptive solutions to address breakdowns and distorted structures within the MG system. Visualized simulation controls interactions, optimizes agile energy transition, and maintains power balance between MG subsystems.

5 Virtual Power Plants (VPPs)

Unlike MGs which have a clear border with other parts of the power distribution system, a VPP is another solution for power system operation and design that is not limited to a small area with determined borders. A VPP is a network of DERs that is controlled as a single entity. Both MGs and VPPs enable utilities to better manage the distribution system and increase its reliability and efficiency. However, VPP concept provides appropriate condition for the integration of various power plants in a large area that may have different scales. For example, integrating wind or solar farms with CHP units creates a larger entity that can provide different services, such as energy or flexibility, for power markets.

Primarily, the idea of VPP was a cloud-based view of some (heterogeneous sorts of) energy resources. This intention would be managed best by DT collaboration. DT integrates, dispatches, and controls DER units in VPP. Like MG, DT provides enhanced software to gather information, sort and analyze related data, and extract valuable information for other managing centers like DSOs. Online analysis of physical structures, their interoperability based on standards, and power flow monitoring and optimization are the main advantages of DTs for VPPs. DTaaS encompasses all requirements of VPPs in their formation and during energy transition process. It provides virtual representations of the various resources within a VPP, such as PV systems, energy storage, and controllable loads, both individually and as a collective. Furthermore, analytical results derived from DT can highlight viable business models and integrate enterprise-level insights to balance financial interests of the VPP stakeholders.

6 Electric Vehicles (EVs)

EVs are becoming increasingly popular. SG subsystems should support the integration of EVs into power distribution systems and control the performance of charging stations, in coordination with the overall load profile. Although the increasing use of EVs is considered in electrification programs, they should also be studied separately, as they impact many management fields, such as traffic control, new road construction, and the development of the SG network. EVs are mobile energy storage units that impact the design and operation of SG systems. Parking lots are also utilized as charging points that integrate EVs into the distribution systems. These parking lot charging stations manage the state of charge (SOC) of EVs based on the power grid requirements and the duration of the cars' presence in the parking spaces.

There are numerous charging stations connected to power distribution system, and these stations may provide various services for EVs, including fast charging. Power consumption of these charging stations is influenced by several factors, such as travel demand across different routes and streets, as well as the SOC of cars. To effectively manage this complex system, a smart distribution system should employ advanced technical software and hardware facilities.

Recently, DT has attracted automotive companies, and some have designed DT for their latest products. DT prompts the automotive industry in subsequent fields:

1. Manufacturing procedures: DT framework offers a comprehensive approach to the planning, design, and construction of EV products. This framework empha-

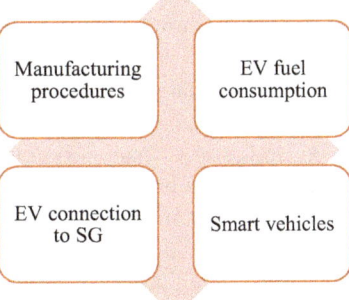

Fig. 5.4 DT advantages for EV

sizes the importance of performance, safety, and efficiency in the EV manufacturing process.

2. EV fuel consumption: DT virtual model of EV batteries represents battery management schemes to ensure their longevity. This DT model can analyze dynamics of the battery system to provide precise information about battery's status, observe energy consumption patterns, and schedule optimal charging strategies.
3. EV connection to SG: Widespread interests in EVs have accelerated the adoption of universal standards and regulations, particularly for EV connections to electricity networks. At this juncture, EV charging stations represent significant energy loads, and DTs can help ensure that they do not compromise the power quality, resiliency, or sustainability of the SG. Public and private charging points can be optimally located, scaled up, and monitored with the assistance of DTs. Furthermore, the DT of a specific EV might exchange information with other DTs deployed within the system, such as the DT of the distribution network.
4. Smart vehicles: DT models could be established to manage the automated driving functions of smart and autonomous vehicles. This includes driver-assistant modules, traffic analysis, extraction of road direction, and long-term path planning (e.g., best battery/fuel usage). Monitoring, maintenance suggestions, and fault diagnosis are other key features that could be best supported by DT of an EV.

The four categories of DT advantages for EVs, as mentioned earlier, are illustrated in Fig. 5.4. In reality, these opportunities are often intertwined within one or more complex structures of DTs for EVs, their components, charging locations, and/or higher management layers (such as DSOs).

7 Energy Markets

Development of energy markets has assisted power systems in numerous cases, specifically improving economic aspects, efficiency, and resiliency of the grid. While energy markets provide a condition in which power generation companies consider their costs and desired profits in setting selling price of electricity, they also incentivize SG toward more stable and efficient economic states.

In traditional power markets which are wholesale markets, participation is generally limited to large entities such as power plants, VPPs, and large loads. However, in SG environment, participation landscape has expanded significantly. Most end-users and power consumers can now actively participate in the market procedure, expressing their energy needs by offering electricity value [45]. Furthermore, other parties within the power system, such as transmission system owners, can now directly engage in the market or utilize results of market competition to inform their decision-making processes.

In SG environments, energy consumption of loads is visible and can be controlled if required. This visibility and controllability of loads enable their participation in wholesale markets, where they can provide ancillary energy services through VPPs. By leveraging VPPs, loads can offer these services at reduced prices compared to traditional methods. Power plants participating in market have to calculate their optimum power prices to cover their costs and also gain their desired profits. Cost term includes various terms such as operational and investment costs. End-users' participation in power market through contracts with VPP reduces prices, as they are willing to reduce their energy consumption in exchange for rewards.

The main factor that determines market prices is the balance between load and generation. During off-peak periods, when load is lower than generation capacity, power plants with lower generation bids have a higher likelihood of being accepted. Conversely, during peak periods, all generation units have a good chance of being accepted, and the market price increases as a result.

DT technology surpasses establishment of local energy markets, where market agents trade energy with other agents based on the perceived value of energy for them. Concepts such as transactive energy systems [46, 47] and peer-to-peer trading [48] are possible in SG.

DT can significantly enhance user experience and simplify interaction between users and the energy market management system. With DT visual illustrations available, immediate consumers' power usage and prices are visible. So, they can react to energy prices and adjust their energy usage to reduce final energy bills. The reaction of consumers to power prices is expressed through their price elasticity, which varies for different types of loads, such as industrial, commercial, or residential loads. DT facilitates swift and convenient market incorporation and helps grid operators with a controlled energy flow.

Chapter 6
Digital Twin for Smart Power Transmission

SG concept advocates for a decentralized structure, aiming to bring electricity generation facilities and consumption end-users as close as possible. This approach allows for the possibility of DERs directly feeding a consumption area without a connection to the main grid. However, centralized large-scale power plants will still be utilized in the foreseeable future, as they provide assured production capabilities. Consequently, long transmission lines will continue to be a part of SG infrastructure.

It is necessary to identify the major features of transmission system, analyze prospects and impacts of DT on performance, design, and monitoring and control of transmission system (as depicted in Fig. 6.1). This way provides DT-based recommendations for optimal or robust transmission network operation and planning under SG scenarios.

1 DT to Improve Transmission Network Performance

DT improves performance of electricity transmission network in the following aspects:

1. Loss reduction of transmission long lines: DT optimizes power flow more effectively, thereby reducing power losses and resulting in significant cost savings.
2. Increasing transmission capacity: DT enables dynamic line rating, which is real-time adjustment of maximum power be transmitted on a line based on environmental conditions such as temperature, wind speed, and solar radiation. DT also increases transmission capacity and utilization of existing lines, avoiding the need for costly and time-consuming network expansion [49].
3. Improved transmission monitoring: DT's platform encompasses enhanced modeling capabilities, enabling smart asset planning in the face of unforeseen

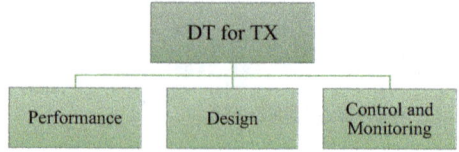

Fig. 6.1 DT for smart transmission

complications and facilitating more efficient powerline inspections. Its visual entities illustrate the virtual behavior of transmission network components, including transformers, circuit breakers, and relays, allowing for real-time monitoring of the status and condition of these critical devices.

4. System-level fault detection: DT enables early detection and diagnosis of faults, as well as rapid isolation and restoration of affected areas. The supplementary insights provided by the DT can notify operators of probable blackouts in the network and automatically trigger autonomous mending and self-healing tactics for the transmission network.
5. Advanced security: DT can also protect transmission networks from cyberattacks by employing encryption, authentication, and intrusion detection systems. DT can suggest the safest course of action based on their effective strategies to ensure the safety and security of transmission infrastructure while enhancing data security.

2 DT for Power Transmission Network Design

SG network topology is affected by reliability, security, and protection requirements in different domains. Conventional viewpoints on transmission networks often suggest the need for additional backup lines or transformers, which are typically utilized only during critical situations, such as unexpected outages of existing infrastructure. However, DT system, equipped with powerful and fast monitoring and control software and hardware, empowers SG system operators to make real-time decisions. This capability allows for the replacement of some backup structures with fast remedial actions. As a result, many of the current preventive actions can be deferred and instead performed through corrective measures.

In the current structure, system operators may not use total capacity of power lines and transformers, as they don't know about real-time conditions of their elements and parameters such as temperature, wind speed, etc. This forces a conservative view not to use their total capacity. Such an operation manner results in the need to expand the extra capacity of lines and transformers. In return, DT enables real-time condition monitoring of transmission lines and transformers. The DT can then utilize dynamic rating of these assets in system operation, ultimately eliminating the need for line or transformer capacity expansion. Such advantages can be drawn in parallel with flexible alternating current transmission system (FACTS) devices, as both technologies improve the flexibility and power transfer capability of transmission networks without the requirement for network expansion.

3 DT for Power Transmission Monitoring and Control

SG system operators utilize wide-area measurement, protection, and control (WAMPAC) solutions in the monitoring, protection, and control of transmission systems. WAMPAC employs sensing, communication, analysis, visualization, and algorithmic techniques to monitor, control, and protect the power transmission network, thereby improving stability and security of such power systems. WAMPAC provides appropriate solutions to overcome the challenges of future power systems, which are expected to contain large renewable resources dispersed over a wide area and connected to load centers using long power transmission lines.

WAMPAC utilizes PMU data to measure voltage, current, frequency, and phase angle of any desired point within the grid. These measurements are captured at a high sampling rate, enabling real-time monitoring and analysis of the grid's dynamic behavior. The collected data is synchronized to a common reference time using Global Positioning System (GPS) signals, ensuring accurate time-stamping and enabling the integration of measurements from multiple locations across the grid. Collected data is used for different applications of state estimation (to estimate state variables of the system, including bus voltages and line currents, and make grid section observable), oscillation detection (to help system operators assess dynamic stability of power system and apply preventive or corrective actions), and voltage stability (by analyzing voltage magnitude and voltage angle information).

WAMPAC has some valuable protection functions. It performs fast and reliable protection actions based on wide-area information. Some of the protection schemes of WAMPAC are explained below:

- Out-of-step protection: In the event of large disturbances, such as faults or sudden load changes, some power generators may lose synchronism with the rest of the power system. When generators fall out of step, it can lead to severe consequences, including equipment damage and potential system collapse. To mitigate these risks, out-of-step protection functions are employed to detect and isolate the affected generators.
- Adaptive protection: Protection schemes may need to be adapted to account for various factors, including load level, penetration of renewable power, and configuration of the network.

In addition to monitoring and protection functions, WAMPAC enables the power system operator to perform control actions, mainly frequency control (to balance load and generation and regulate system frequency in the allowed range), voltage control (voltage control modules action and regulate voltage profile by helping of reactive power generation of generators, performance improvement of FACTS and load control devices), and power flow control (phase shifters change flow of power in network using real-time data of lines' flow and voltage angles recorded by PMUs).

Although the performance of DT appears to be close to WAMPAC, it suggests supplementary modules to fully realize online and offline applications of WAMPAC. DT has the capability to monitor multiple non-parallel levels and perform analysis across different layers of transmission network, including physical layer, automation and control network, as well as application and security levels.

Chapter 7
Digital Twin for Future Smart Grid

Power systems are vast geographical networks that require secure and expeditious data exchange among their diverse subsystems and components. The emergence of SG has revolutionized power systems by seamlessly integrating IDT techniques. This integration combines SG with digitalization and has provided necessary conditions for efficient data transmission, information management, analysis, visualization, and decision-making processes.

Recent and widespread electrification strategies across various energy sectors, such as transportation and heating demand, have been driven by several motivations, including environmental concerns and economic factors. This electrification process has significantly impacted administration of electricity, introducing presence of prosumers with diverse production and consumption patterns. Electrification plans have particularly incorporated new loads, such as energy storage systems and electric vehicles, which have introduced substantial charging demands.

Digitalization will be a pivotal pillar in the future power grid's journey towards a SG ecosystem and latest digital technologies have proved beyond satisfactory effects in industries. Without a doubt, SG subsystems are accommodating recent technologies such as IoT, cloud computing, ML methods, and AI techniques. However, it is enormously advantageous when the proficiency of these technologies is federated to form a stronger configuration, known as a digital twin. The succeeding DT embraces trending digital technologies to aggregate their technical expertise and provide dominant skills for SG.

DT outreaches mere digitization in SG subsystems. Bear in mind that DT is a proactive tool that empowers SG in supervising the immense and complex power system, which serves a wide range of energy needs and end-users. Cognitive-DT and networked-DT enhances SG applications that are receptive to advanced DT solutions, leading to increased system stability and efficiency.

Figure 7.1 illustrates a glance at two technical paths that recently happened in parallel, the former depicts DT evolution while the latter shows actual

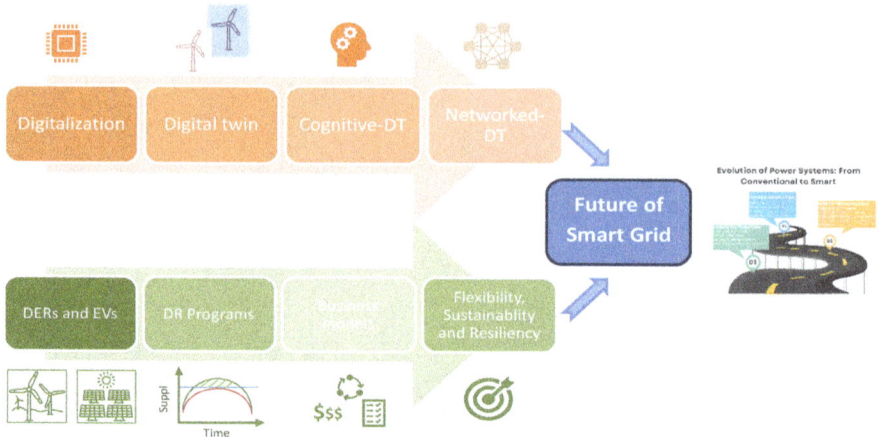

Fig. 7.1 Future of smart grid

enhancements in power system's modules. Content of this book discussed these two trends and their hand in hand collaboration towards future of SG.

Experts are also mindful of the fact that current power networks, which are predominantly based on alternating current (AC) systems, will turn into hybrid (AC–DC) power systems in the near future. This transition is due to the increasing use of renewable resources, which produce power in DC form, and the replacement of AC power consumption technologies with DC ones, such as brushless DC (BLDC) motors replacing AC electric motors. Therefore, the future of SGs will be a combination of AC and DC systems separated by converter devices. Many power-electronic devices, along with operating variables in both AC and DC sections, have to be controlled using software and hardware facilities.

Flexibility of SG is another significant topic of study. Essential requirements of flexibility factors dictate the integration of volatile DERs into SG. As penetration of wind and solar resources in power systems increases, DT technology provides system operators with valuable informative resources. These resources enable operators to effectively manage uncertainty and variability inherent in renewable energy sources. DT also strengthens the balancing market, which is a widely adopted mechanism for addressing these challenges. To optimize its effectiveness, the balancing market should be operated in near real-time, closely mirroring actual system conditions. Currently, individual DERs are not permitted to directly participate in the balancing market. However, DERs can contribute to providing flexibility services through VPPs at competitive prices.

Despite benefits of SG programs, their feasibility should be carefully considered when integrating them into the current power system. Ambitious targets of electrification, coupled with the goal of exclusively relying on renewable electrical production, compel policymakers to encourage more investments in technological innovations. This is necessary to make SG more efficient and sustainable.

7 Digital Twin for Future Smart Grid

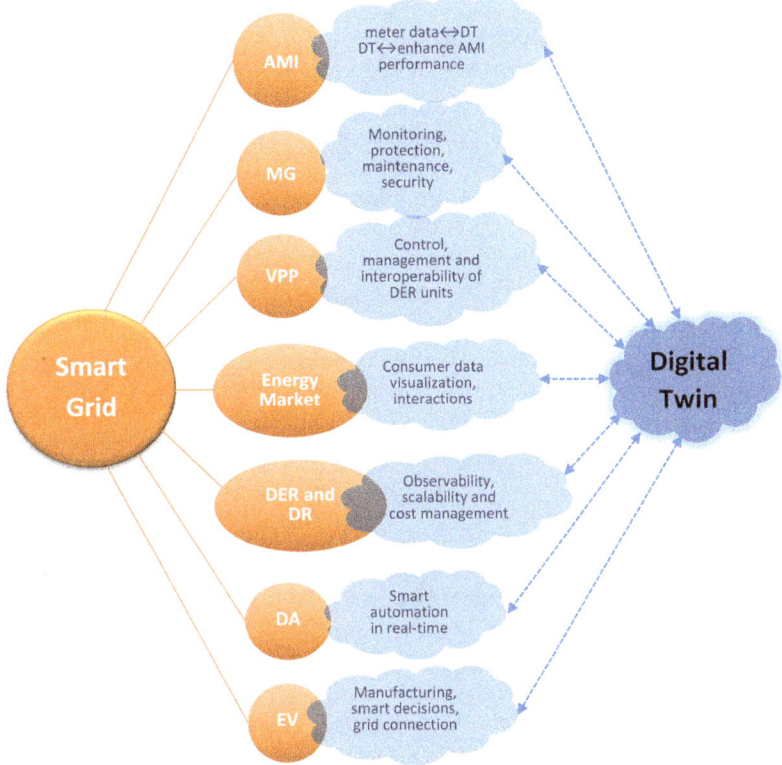

Fig. 7.2 DT effects on SG main subsystems

Figure 7.2 illustrates DT effects and improvements on SG main subsystems. Other DT opportunities for SG in the near future are titled as follows:

- DT delivers immersive experiences for engineers when AR and VR intuitive modules are appointed. This could be a sustainable solution for asset management, particularly in the context of predictive and preventive maintenance.
- AI is another upheaval technology that involves learning methods in DT. This is designated as the "cognitive digital twin (CDT)," which performs self-learning of a subsystem through metadata analysis and renders optimal strategies for future network developments.
- The cross-sectoral modeling and simulation of DTs can help reduce outage risk in a local distribution network by addressing abnormal conditions.
- DT illustration of feeders helps visual inspection of faults, their principal causes, and restoration solutions.
- To mitigate the negative impacts of EV charging stations on power network, interaction of several DTs is recommended. These DTs should include DT of EVs, charging points, charging stations, and DSO.

- DT approach facilitates the convergence of common regulation for multivendor devices, enabling unified performance monitoring and operational maintenance in a system with a networked-DT structure.
- DT evolution has led to disruptive business models. It proposes innovative energy services and customizes advice for different SG applications.
- Collaborative and problem-solving interactions within a networked DT system can have a beneficial effect on EMS, particularly in the areas of cost reduction and outage management.
- Social approaches for peak load shaving strategies have gained importance in recent years. DT can assist policymakers in this regard by providing a global view of the power system. DT runs to pre-study effects of different incentives and help decide on alternative DR programs in local power networks.
- Educational tools have been revolutionized by VR and AR technologies, which are key components within the DT architecture. Interactive nature of DT is best for skill training and learning goals in all SG domains.

In general, conducting viability assessment requirements are similar or close in most SG subsystems. This suggests that once a DT architecture is designed and assembled, the framework can be customized and applied to other SG modules. In practice, we need standard specifications with congruent interfaces and collaborative efforts from all stakeholders to unify component models. These prerequisites are essential for exchanging dynamic device models and enabling real and effective interaction between some DTs. Moreover, a sustainable business study is necessary to realize the full potential of DTs for electric companies, not just as intermediaries, but as a key driver for industrial organizations.

Ultimately, the undeniable opportunities presented by DTs for SG power systems will enhance even the most obscure prospects of SG subsystems. Although barriers, primarily cost and a lack of widespread knowledge, hinder the rapid penetration of DT technology in industry, the subject remains of great interest in scientific and research centers.

Looking ahead, we envision that DT technology is not merely a futuristic concept but rather a practical solution that will contribute to various SG ecosystems in near future. DT has the potential to optimize design and operation of future power structures, ultimately leading to the development of more energy-efficient and sustainable aspects of human life.

Chapter 8
Conclusions

Enhanced algorithms and faster processing units in the ever-increasing number of recent digital techniques have had a significant impact on real life. Trending digital twin (DT) technology engages and convenes various digital techniques, enabling swift and online data analysis, models, and simulations for industries.

SG is a multidisciplinary, multi-domain system that necessitates pervasive deployment of information and communication technologies (ICTs). The upcoming SG subsystems are anticipated to be intelligent and sustainable, with a significant proportion of renewable energy sources such as wind and solar. Since DT extracts and visualizes credible information out of massive data, it facilitates and consolidates SG management with its innovative and conductive suggestions. Overall, DT will be one of the best tools to reaching SG goals.

In this book, the concept of DTs and their underlying architecture and enabling technologies were overviewed. The discussion then delved into the various applications of DTs, benefits they offer, and challenges faced by industry in their adoption. Subsequently, the text proposed a structure for SG and examined how the process of digitalization has impacted different subsystems within SG. The abovementioned topics were then combined to capture DT potentials to improve the overall efficiency of SG. In the preceding four chapters, perspectives of DT for future electricity generation, distribution, and transmission were specifically studied. Finally, effects of DT on the approaching SG in the near future were evaluated.

DT has been widely recognized as a promising solution for the power industry. The successful development of a DT for a specific system within the SG can deliver informative and practical results based on real-time data. This unparalleled talent of DT grants more resiliency, reliability, and flexibility to the energy sector.

References

1. Tao, F., Zhang, M., Nee, A.Y.C.: Digital Twin Driven Smart Manufacturing. Academic Press, New York (2019)
2. Jones, D., Snider, C., Nassehi, A., Yon, J., Hicks, B.: Characterising the digital twin: a systematic literature review. CIRP J. Manuf. Sci. Technol. **29**, 36–52 (2020)
3. Analytics, I.: How the world's 250 Digital twins compare? Same, same but different (2020)
4. Glaessgen, E., Stargel, D.: The digital twin paradigm for future NASA and U.S. Air Force Vehicles. In: 53rd AIAA/ASME/ASCE/AHS/ASC Structures, Structural Dynamics and Materials Conference
20th AIAA/ASME/AHS Adaptive Structures Conference
14th AIAA. American Institute of Aeronautics and Astronautics, Honolulu (2012)
5. Lee, J., Lapira, E., Bagheri, B., Kao, H.: Recent advances and trends in predictive manufacturing systems in big data environment. Manuf. Lett. **1**, 38–41 (2013)
6. Grieves, M., Vickers, J.: Digital twin: mitigating unpredictable, undesirable emergent behavior in complex systems. In: Kahlen, F.-J., Flumerfelt, S., Alves, A. (eds.) Transdisciplinary Perspectives on Complex Systems, pp. 85–113. Springer International Publishing, Cham (2017)
7. Söderberg, R., Wärmefjord, K., Carlson, J.S., Lindkvist, L.: Toward a digital twin for real-time geometry assurance in individualized production. CIRP Ann. **66**, 137–140 (2017)
8. Zhang, M., Sui, F., Liu, A., Tao, F., Nee, A.Y.C.: Digital twin driven smart product design framework. In: Digital Twin Driven Smart Design, pp. 3–32. Elsevier, Amsterdam (2020)
9. O'Connell, E., O'Brien, W., Bhattacharya, M., Moore, D., Penica, M.: Digital twins: enabling interoperability in smart manufacturing networks. Telecom. **4**(2), 265–278 (2023)
10. Bazaz, S.M., Lohtander, M., Varis, J.: 5-Dimensional definition for a manufacturing digital twin. Procedia Manuf. **38**, 1705–1712 (2019)
11. Qi, Q., Tao, F., Hu, T., Anwer, N., Liu, A., Wei, Y., Wang, L., Nee, A.Y.: Enabling technologies and tools for digital twin. J. Manuf. Syst. **58**, 3–21 (2021)
12. Stark, R., Fresemann, C., Lindow, K.: Development and operation of digital twins for technical systems and services. CIRP Ann. **68**, 129–132 (2019)
13. Lim, K.Y.H., Zheng, P., Chen, C.-H.: A state-of-the-art survey of digital twin: techniques, engineering product lifecycle management and business innovation perspectives. J. Intell. Manuf. **31**, 1313–1337 (2020). https://doi.org/10.1007/s10845-019-01512-w
14. Digital twin of material handling chain—ABB Ability Stockyard Management System (mining operations and production management | ABB). https://new.abb.com
15. GE digital twin. http://www.ge.com/digital/sites/default/files/download_assets/Digital-Twin-for-the-digital-power-plant-.pdf
16. Digital twin. https://new.siemens.com/global/en/company/stories/research-technologies/digitaltwin/digital-twin.html

17. Quick insight brief: digital twin activities at EPRI. https://www.epri.com/research/products/000000003002020014 (2020)
18. Kostenko, D., Kudryashov, N., Maystrishin, M., Onufriev, V., Potekhin, V., Vasiliev, A.: Digital twin applications: diagnostics, optimisation and prediction. In: Annals of DAAAM & Proceedings, vol. 29 (2018)
19. Javaid, M., Haleem, A., Suman, R.: Digital twin applications toward industry 4.0: a review. Cogn. Robotics. **3**, 71–92 (2023)
20. Botín-Sanabria, D.M., Mihaita, A.-S., Peimbert-García, R.E., Ramírez-Moreno, M.A., Ramírez-Mendoza, R.A., Lozoya-Santos, J.d.J.: Digital twin technology challenges and applications: a comprehensive review. Rem. Sens. **14**, 1335 (2022)
21. Buchholz, B.M., Styczynski, Z.: Smart Grids-Fundamentals and Technologies in Electricity Networks. Springer, Berlin (2014)
22. Hashemi, S.M., Alizadeh, B., Sheibani, M., Fallahi, F.: Assessing potential of renewable energy sources in Iran through practical and analytical data. In: 2023 13th Smart Grid Conference (SGC), pp. 1–8. IEEE, Piscataway (2023)
23. Hashemi, S.M., Tabarzadi, M., Fallahi, F., Kalhori, M.R.N., Abdollahzadeh, D., Qadrdan, M.: Water and emission constrained generation expansion planning for Iran power system. Energy. **288**, 129821 (2024)
24. List of most powerful wind turbines. https://en.wikipedia.org/w/index.php?title=List_of_most_powerful_wind_turbines&oldid=1148223960#cite_note-twittergamesa11200-2 (2023)
25. Mohan, N., Undeland, T.M., Robbins, W.P.: Power Electronics: Converters, Applications, and Design. John Wiley & Sons, New York (2003)
26. Kia, M., Shafiekhani, M., Arasteh, H., Hashemi, S.M., Shafie-khah, M., Catalão, J.P.S.: Short-term operation of microgrids with thermal and electrical loads under different uncertainties using information gap decision theory. Energy. **208**, 118418 (2020)
27. Zheng, Y., Jenkins, B.M., Kornbluth, K., Kendall, A., Træholt, C.: Optimal design and operating strategies for a biomass-fueled combined heat and power system with energy storage. Energy. **155**, 620–629 (2018)
28. Rawat, A.K., Shukla, A.K., Sharma, M., Singh, A.K.: Geothermal energy-based combined cooling heating and power system. J. Phys. Conf. Ser. **2178**, 012040 (2022)
29. Sharma, S., Basu, S., Shetti, N.P., Kamali, M., Walvekar, P., Aminabhavi, T.M.: Waste-to-energy nexus: a sustainable development. Environ. Pollut. **267**, 115501 (2020)
30. Pandey, V.C., Singh, K., Singh, J.S., Kumar, A., Singh, B., Singh, R.P.: Jatropha curcas: a potential biofuel plant for sustainable environmental development. Renew. Sust. Energy Rev. **16**, 2870–2883 (2012)
31. Rodionova, M.V., Poudyal, R.S., Tiwari, I., Voloshin, R.A., Zharmukhamedov, S.K., Nam, H.G., Zayadan, B.K., Bruce, B.D., Hou, H.J., Allakhverdiev, S.I.: Biofuel production: challenges and opportunities. Int. J. Hydrog. Energy. **42**, 8450–8461 (2017)
32. DiPippo, R.: Geothermal Power Plants: Principles, Applications, Case Studies and Environmental Impact. Butterworth-Heinemann, Oxford (2012)
33. Kanakasabapathy, P., Swarup, K.S.: Bidding strategy for pumped-storage plant in pool-based electricity market. Energy Convers. Manag. **51**, 572–579 (2010)
34. Budt, M., Wolf, D., Span, R., Yan, J.: A review on compressed air energy storage: basic principles, past milestones and recent developments. Appl. Energy. **170**, 250–268 (2016)
35. Zhang, X., Qin, C.C., Xu, Y., Li, W., Zhou, X., Li, R., Huang, Y., Chen, H.: Integration of small-scale compressed air energy storage with wind generation for flexible household power supply. J. Energy Storage. **37**, 102430 (2021)
36. Arasteh, H., Moslemi, N., Hashemi, S.M.: Demand response measurement and verification approaches: analyses and guidelines. In: Industrial Demand Response: Methods, Best Practices, Case Studies, and Applications, p. 133. IET, London (2022)
37. Hashemi, S.M., Vahidinasab, V.: Energy management systems for microgrids. In: Microgrids: Advances in Operation, Control, and Protection, pp. 61–95. University of Salford Manchester, Salford (2021)

References

38. Hashemi, S.M., Vahidinasab, V., Ghazizadeh, M.S., Aghaei, J.: Reliability-oriented DG allocation in radial microgrids equipped with smart consumer switching capability. In: 2019 International Conference on Smart Energy Systems and Technologies (SEST), pp. 1–6. IEEE, Piscataway (2019)
39. Hashemi, S.M., Vahidinasab, V., Ghazizadeh, M.S., Aghaei, J.: Security constrained operation of radial micro grids based on the loads' vulnerabilities and flexibilities. Tabriz J. Electr. Eng. **50**, 1441–1453 (2020)
40. Hashemi, S.M., Vahidinasab, V., Ghazizadeh, M.S., Aghaei, J.: Valuing consumer participation in security enhancement of microgrids. IET Gener. Transm. Distrib. **13**, 595–602 (2019)
41. Hashemi, S.M., Vahidinasab, V., Ghazizadeh, M., Aghaei, J.: Load control mechanism for operation of microgrids in contingency state. IET Gener. Transm. Distrib. **14**, 5407–5417 (2020)
42. Hashemi, S.M., Arasteh, H., Shafiekhani, M., Kia, M., Guerrero, J.M.: Multi-objective operation of microgrids based on electrical and thermal flexibility metrics using the NNC and IGDT methods. Int. J. Electr. Power Energy Syst. **144**, 108617 (2023)
43. Arasteh, H., Bahramara, S., Kaheh, Z., Hashemi, S.M., Vahidinasab, V., Siano, P., Sepasian, M.S.: A system-of-systems planning platform for enabling flexibility provision at distribution level. In: Flexibility in Electric Power Distribution Networks, pp. 41–65. CRC Press, Boca Raton (2021)
44. Wu, Y., Guerrero, J.M., Wu, Y., Bazmohammadi, N., Vasquez, J.C., Cabrera, A.J., Lu, N.: Digital twins for microgrids: opening a new dimension in the power system. IEEE Power Energy Mag. **22**, 35–42 (2024)
45. Troncia, M., Galici, M., Mureddu, M., Ghiani, E., Pilo, F.: Distributed ledger technologies for peer-to-peer local markets in distribution networks. Energies. **12**, 3249 (2019)
46. Ambrosio, R.: Transactive energy systems. IEEE Electrif. Mag. **4**, 4–7 (2016)
47. Huang, Q., Amin, W., Umer, K., Gooi, H.B., Eddy, F.Y.S., Afzal, M., Shahzadi, M., Khan, A.A., Ahmad, S.A.: A review of transactive energy systems: concept and implementation. Energy Rep. **7**, 7804–7824 (2021)
48. Tushar, W., Saha, T.K., Yuen, C., Smith, D., Poor, H.V.: Peer-to-peer trading in electricity networks: an overview. IEEE Trans. Smart Grid. **11**, 3185–3200 (2020)
49. Butt, O.M., Zulqarnain, M., Butt, T.M.: Recent advancement in smart grid technology: future prospects in the electrical power network. Ain Shams Eng. J. **12**, 687–695 (2021)

Index

B
Building technologies, 5–7, 12, 13

D
Design, 2, 5–7, 9–12, 23, 35, 36, 39
Digitalization, 18–19, 43, 47
Digital twins (DTs), 1, 9, 15, 21, 29, 39, 43, 47
Distribution system (DS), 29, 30, 33–36
DT advantages, 24, 37
DT applications, 6, 7
DT benefits, 11–13
DT challenges, 11–13

E
Energy storage, 23, 27–29, 33, 36, 43

F
Future SG, 43–46

M
Monitoring and control, 29, 30, 34, 39–42
Multi-energy generation system, 25, 27

P
Performance, 3, 6, 9, 10, 15, 21–23, 25–27, 31, 36, 37, 39, 41, 42

R
Renewable generation system, 29

S
Smart grid structure (SG structure), 16–18, 33
Smart power generation, 22
Smart power transmission, 39–42

GPSR Compliance

The European Union's (EU) General Product Safety Regulation (GPSR) is a set of rules that requires consumer products to be safe and our obligations to ensure this.

If you have any concerns about our products, you can contact us on

ProductSafety@springernature.com

In case Publisher is established outside the EU, the EU authorized representative is:

Springer Nature Customer Service Center GmbH
Europaplatz 3
69115 Heidelberg, Germany

www.ingramcontent.com/pod-product-compliance
Lightning Source LLC
Chambersburg PA
CBHW072122150625
28252CB00002B/29